Quantum Art and Uncertainty

Quantum Art and Uncertainty

Paul Thomas

intellect Bristol, UK / Chicago, USA

First published in the UK in 2018 by
Intellect, The Mill, Parnall Road, Fishponds, Bristol, BS16 3JG, UK

First published in the USA in 2018 by
Intellect, The University of Chicago Press, 1427 E. 60th Street, Chicago, IL 60637, USA

A catalogue record for this book is available from the British Library.

Copy-editor: MPS Technologies
Cover designer: Aleksandra Szumlas and Paul Thomas
Index: Lyn Greenwood
Production manager: Mareike Wehner
Typesetting: Contentra Technologies

Print ISBN: 978-1-78320-901-9
ePDF ISBN: 978-1-78320-903-3
ePUB ISBN: 978-1-78320-902-6

This is a peer-reviewed publication.

Contents

Acknowledgments

I would like to acknowledge the support of Thomas Retter for his reflections, challenges, comments and elucidations and the support of my collaborator Kevin Raxworthy for all his work in developing the technical visualization of the quantum phenomena. I would also like to draw attention to the work of Professor Andrea Morello and Dr Juan Pablo Dehollain for their support with my artwork *Quantum Consciousness* based on their research in Quantum Computing. Thanks goes to Emma Crott and special thanks goes to Timothy Morten for contributing the foreword for this book.

Finally, I would like to thank my wife Christina Newberry for putting up with me endlessly talking about quantum uncertainties.

Paul Thomas, 2018

Foreword

Between Nothingness and Eternity

Timothy Morton
Rice University

What is life? It's a question answered most interestingly by physicists, not biologists. It's a curious symptom of biology that it cannot question its initial starting assumption, namely the 'life' of which it is the science. Debates about the origins of life are notoriously becalmed in an ocean of doubt concerning whether metabolism or replication is the driver. This means that evolution theory, for example, can't actually explain the origin of a single-celled organism, the most fascinating aspect of the whole thing. Evolution theory can explain what happens next – all lifeforms unintentionally act like human pigeon breeders, exerting adaptive pressure on them from all sides. But the starting position remains obscure.

Two physicists have offered interesting explanations of life. One is a cosmologist, the other a quantum theorist. But the cosmologist deals with photons in his explanation, which makes him a quantum theorist by default. The quantum theorist is Erwin Schrödinger, whose *What Is Life?* has recently undergone a refreshment of study and discussion.

The cosmologist, a young physicist named Jeremy England, has recently argued that if you just beam photons at atoms for long enough time, life will emerge. This is simple and amazing, because it implies that there really isn't all that much to life (as opposed to non-life) as one might have thought. It's a readily-available aspect of our reality. It's all about patterns.

Schrödinger's *What Is Life?* is fascinating, because it argues that I wouldn't be stable enough to write this sentence if it weren't for millions of tiny quantum events taking place in my body. I would dissolve into a cloud of powder. My very solidity depends on how particles can suddenly change, become entangled, blur into one another. These properties bestow upon me what Schrödinger calls *negentropy*, a temporary suspension of the normal rules whereby the arrow of time moves only in one direction and things tend towards collapsing. Life is a little pocket of stability in the universe.

But what kind of stability? A dynamic, *metastable* state that isn't solid or permanent according to our ideas of such things. Life isn't total presence, always consistent and smooth. It's like one of Paul Thomas's extraordinary artworks, subtle and trembling.

Where do we locate this trembling in a metaphysical sense? We locate it between two kinds of death. Life is not the opposite of death, but a strange quivering of the needle between two different types of being dead. At one end of the spectrum we have total nonexistence. At the other end, we have total mechanical repetition, perpetual motion, the machination that Freud called the *death drive*. If the needle stops quivering between these poles and slips towards the pole of total nonexistence, the living being begins to dissolve, to drift into powder, to become the chemicals that composed it. If the needle stops quivering and is attracted to the pole of mechanical repetition, the living being becomes a relentless force, like a drill or a tornado.

We live in a world where there is way too much attraction towards the mechanical pole, in the name of warding off the pole of total nonexistence. Agriculture as formulated in Mesopotamia and elsewhere in about 10,000 BC eventually gave rise to industry in order to keep its mechanically repetitive logistics going, and the end result, 200 years later, is mass extinction, politely called global warming, which in turn is politely called climate change. It's the death drive all right. Humans have been weaponized. They have become a geophysical force on a planetary scale. Despite our individual intentions, and despite the actual statistical meaninglessness of our individual actions (don't feel guilty, ecological awareness isn't about that at all), humans as a massive collective force have become like the asteroid that wiped out the dinosaurs 65 million years ago, the asteroid that was the cause of the most recent mass extinction, the fifth one. Think about that. You are part of an asteroid hurtling towards earth, metaphorically speaking. As you read this, your human beingness is hurtling towards earth with massive destructive momentum.

Art suspends and postpones this momentum. One keen reader of poetry once declared that poetry was the postponement of the end of the world. One more quivering, for old times' sake. One more tremble. One more flicker of the needle, defying the magnetic attraction of the weaponizing mechanical death drive.

Art that works on the trembling, fragile edge of quantum phenomena is particularly suspended between the two deaths. Paul Thomas does what artists do: he resists collapsing into oblivion (the art and culture equivalent of total nonexistence), and he resists falling into mechanical repetition (the dull round of churning out 'stuff'). Thomas's work is alive, but not in your granddaddy's sense of being opposed to death, or for that matter, Ray Kurzweil's sense of wanting to cheat death by having his body frozen until humans have conquered death. Such fantasies are part of the logistical structure that is now destroying earth. We should learn how to dissolve a little bit. A little bit of total nonexistence wouldn't be so bad right now.

Life as quivering between nothingness and eternity isn't so easy to distinguish from the state we call *undeath*. This ghostly flicker is lively but not *definitely* alive, as we've seen. So thinking about lifeforms this way means that we allow them to be uncanny, hard to discriminate, hard to build walls around, inviting otherness, open to strangers, with all that this entails, as in the word *host*. This word comes from a Latin word (*hostis*) that can mean both *friend* and *enemy*. Exactly: real life is very uncertain, and a good thing too. The maniacal pursuit of certainty becomes more and more efficient at destroying its environment. There is some kind of link between the art of the quantum and the caring for the biosphere we call ecological awareness and politics. Quantum states are notoriously delicate, capable of collapsing into regular old 'classical' ones if you interfere with them too much. Likewise, beauty can be ruined if you think about it too hard or try to isolate its active ingredient in the work you are admiring. This fragility is immensely powerful, because it disturbs and challenges rigid binary ways of thinking and acting.

The power of uncertainty. That's what this book is all about.

It's a great honour to introduce this book to you. Paul Thomas is a brilliant observer of things, a fantastic decelerator who interrupts the normal speed of contemporary life. I once took a walk down a street with him that seemed to last for thousands of years because of the amazingly careful observation of every flake of paint, every crack between every building. Thomas is an observer of the in-between, as I've been arguing, the in-between that so often gets left out of the nice binary picture. But we all know that total nonexistence is impossible – plenty of philosophers, all the way back to Aristotle, have argued that there simply can't be a void, because it would be impossible to measure or to traverse at all – if it had parts, in other words if it could be measured, it wouldn't be a void. And we all know that perpetual motion is impossible. These poles are really tendencies within a continuum that is characterized entirely by 'in-between-ness', only now we need a better term because we're no longer talking strictly about something in between two genuine poles. Perhaps we need to say *spectrality*? I'm uncertain.

With the non-judgemental care characteristic of a good artist, Thomas has assembled a remarkable series of explorations here, explorations of every type of quantum phenomenon as they pertain to art. What's particularly wonderful is how this book reconfigures what we think we know about the past, like all good art does. History is restructured. Turner, Bacon, Malevich, Coldstream, Cubism, Rauschenberg … history is in here, transmogrified.

Towards the very end of the book, Thomas asks, 'Is there a rupture that can occur in the human psyche that will enable us to see the atomic truth of our ongoing coexistence with the material world of which we are constituted but which we treat as other?' Exactly. But this rupture is best thought as a gentle, and therefore ultimately threatening, quivering slit, a letting-vibrate of things that vibrate all by themselves without being mechanically pushed. Paul Thomas's study is a way to begin to think how to let this happen.

Introduction

The Quantum Atmospherics of Consciousness

This book builds on experimental and transdisciplinary ideas I first broached in my previous publication, *Nanoart: The Immateriality of Art*. In *Nanoart*, I sought to articulate a series of synchronistic relationships that reveal artistic practice as integral to giving form and order to our sense of the physical world. In this current publication, I continue to inquire into object-oriented ontologies (OOO), speculative realism and new materialism[1] – all topics of discussion in my previous book – but in *Quantum Art and Uncertainty* these ideas are all brought to bear on the problem of quantum phenomena and the epistemological challenge it poses to creative and scientific inquiry. I adopt an experimental discourse that challenges perspectives by exploring subconscious and conscious fragments that have been, and are currently, influencing the cultural discourse of our time. This book, via the act of experimental writing, reveals quanta that reflect, and are analogous to, the complex problem of making art that questions reality in the age of uncertainty when dealing with an immanence or speculation. This book does not intend to be a historical resource on quantum theory, uncertainty and art but rather a resource for the broad ideas in this field and how they can be engaged within the visual arts. As an artist, it is the peculiar nature of ideas and their poetic resonance for a particular vision, version, image, expression or sense of the world that challenges the creation of a new visual language. The idea does not have to be confirmed as true but has the potential to make work that enables a reconfiguration or rupture in how the object world is defined. For example, K. Eric Drexler's 1986 publication *Engines of Creation: The Coming Era of Nanotechnology* gave a public face to nanotechnology, forecasting a potential problem which he called 'gray goo' (1986: 146) that could de-territorialize all matter. Drexler stated that '[t]he gray goo threat makes one thing perfectly clear: we cannot afford certain kinds of accidents with replicating assemblers' (1986: 147). This is an example where speculative thinking created by science is refuted by science, yet the thought still has strong social and cultural agency. It is a strange kind of poetics at the heart of all matter that makes our comprehension of matter of such great interest to artists. The concept 'gray goo' does not need to be true; it is a provocation, a thought experiment that garners alternative problem finding for artists.

In this publication there is no expert authoritative voice, there is no hierarchy of knowledge, because we know through speculative realism that human comprehension is limited and the only possibilities are arbitrary and contingent speculations. Unlike nanotechnology, it is hard to pin down the exact moment that the real effects of quantum mechanics[2] first took hold on creative imagination, and therefore where the real synergies lie between quantum theory and creativity. 'Quantum'[3] has been around for over a hundred years as a theoretical construct, but only in the twenty-first century, in light of the emerging agency of quantum computing, is it truly coming into its own.

My interest in quantum theory was initially piqued during my first years at art school – even though I didn't fully appreciate it at the time. I asked a question in the process of drawing a specific object in a still-life arrangement as to its spatial whereabouts; where did this object exist in space? I proposed that if I were to put a mirror in a given position within the room, the object would seemingly reappear – and would do so over and again for every mirror that duplicated the object. The object, I thought, must then be omnipresent, for a part of it could be replicated ad infinitum, limited only by the number of mirrors the room could hold. I now realize that at that time I was speculating on the physical conditions and mechanics of photons and the quantum materiality of the everyday. The light that made the still-life object visible, that brought the object into existence via sight, also transported the object to all other possible locations. The still-life object, the very icon of stasis and empirical stability, was here, I realized, entangled in a condition of ontological uncertainty.

My initial interest as an artist arriving in Australia in 1977 was in spatiality – an 'immigrants' narrative but also as an explorer's one (in the aesthetic exploration of new spatialities) searching and forming a series of speculations of my space and time. This led me to go through a scaling down process towards the infinitely small of nanotechnology and finally to the quantum.

It was the inescapable problem of uncertainty that spurred my interest in quantum physics and in particular the photonic work of Richard Feynman.[4] At the time, Feynman was the arbiter of quantum theory in popular science. He seemed to me to distil the very problem I had encountered. Feynman states on uncertainty in a 1956 talk given at the Caltech YMCA Lunch Forum:

> I think that when we know that we actually do live in uncertainty, then we ought to admit it; it is of great value to realize that we do not know the answers to different questions. This attitude of mind – this attitude of uncertainty – is vital to the scientist, and it is this attitude of mind which the student must first acquire. It becomes a habit of thought. Once acquired, one cannot retreat from it any more.
>
> (Feynman 1956)

Taking the 1927 *Fifth Solvay International Conference on Electrons and Photons* as a touchstone, a significant point in time, this book will explore concepts of art and physics and their contributions to our image of the world. As a catalyst for the reunification of art, science and culture, this conference engendered a discourse of uncertainty and probability in the world that infected all disciplines of knowledge and inquiry, not simply physics. The 1927 Solvay conference aligned, for the first time, the creative and epistemological problem of the imaginary and the unknown with the imperceptible world of the quantum, coalescing into a global sense of 'becoming'.[5]

In this book, concepts of quantum phenomena are dealt with through the perceptions of an artist as to quantum social and cultural impact. Of great importance is the way in which quantum has been historically, and is now currently, affecting our understanding of the everyday world that we constitute. To achieve this, we need to first create some definitions, to lay down the parameters of what is to be discussed. The Newtonian world was built up by making rules to explain everything we saw happening within the world. Physics started questioning the Newtonian laws that did not correlate with new theoretical experiments being formulated. These new sets of interrogations needed a different set of rules applied to the infinitely small that did not fit with our classical way of thinking and seeing – they were quantum.

To explain the infinitely small as a new way of thinking, a new set of rules was applied via a fundamental classical experiment; the 'double slit experiment'.[6] Think of a fence and in front of the fence is a wall with two doors. We throw an apple through one door and then another through the other door. After a while we find we have two clumps of grouped apples stuck in the fence. Now, in the next quantum experiment, the apples are analogous to small particles going through the same two open doors and hitting a long sensor rather than a fence. What we find is the small particles counter-intuitively form clusters of stripe patterns on the sensor. If you decide to close one of the doors and throw small particles through the other, you would form a cluster on the sensor, but if you open the other door and repeated throwing the small particles through the one door what would happen? You think you could guarantee the small particle going through that one door again forming a cluster pattern. This would seem obvious in a classical world, but in a quantum world you would be wrong. Quantum physics has to give up on the statement that each small particle goes through just one door and therefore has to create a new set of rules for explaining these phenomena. Quantum physics says that if the small particle is not in position A, B or C then it must be in all possible positions at the same time; what is called a superposition.

Quantum physics is fundamentally confusing and therefore not to be known. It appears to be counter-intuitive, even for the eminent physicist Richard Feynman who said in a televised lecture that 'nobody understands quantum mechanics'.[7] We have to use alternative ways of exploring these phenomena that can make sense of this world. It relies on mathematical probabilities to experience what can't be defined, to intuit what can't be comprehended. It is not to imagine what is not there but 'our imagination is stretched to the utmost just to comprehend those things which are there' (R. Feyman 1964).

Key themes in this book are: the wave/particle duality, superposition, quantum entanglement, atomic spin, material agency and the ontology of uncertainty. These themes are all brought to bear upon, and through, the need for an injection of new experimentalism in the ways in which we give image to the world so as to inculcate a consciousness of the quantum. In the chapters that follow the problem of uncertainty is historicized within two distinct systems of thought: perspectivalist realism and quantum mechanics. These two different systems have each coerced new imaging tactics from creative fields, that is to say they have broken from tradition so as to instil a new paradigm of knowledge and thinking befitting the possibilities of material agency.

Each chapter in this book hones in on a unique property or dimension or aspect of quantum phenomena and the impact of uncertainty. I offer a brief overview of these streams below that focus on the metaphorical cloud, diagrams, the spin, the swerve and graphene.

Chapter 1: The Swerve – explores the swerve of Epicurus, Democritus and Lucretius[8] as an endorsement in quantum mechanics laying the platform for new forms of exploring, expressing and thinking of materiality and agency within the arts. This chapter reflects upon the legacy and reality of Heisenberg's 'uncertainty principle', which he introduced at the 1927 Solvay conference. The surrealist artists of the 1930s are put forward as a correlate to Heisenberg's postulations – most notably Roberto Matta and his attempts to find a new visual language that embraced the concepts that physicists, like Heisenberg, were uncovering. The turbulence, flux, eddies and flows demonstrate an ongoing foundational investigation into the connections between art and physics. The case is made that artists, too, are adroit at intuiting the realities of quantum phenomena and are champions of uncertainty. This chapter weaves through figures such as Paul Cézanne, Martin Creed and Alberto Giacometti, and makes the case that all these practitioners are in essence focusing (in their work) on exploring the phenomenon of uncertainty. By way of the nature of their trade, artists are familiar with, and masterful navigators of, the visual unknown and the void.

In Chapter 2: The Diagram – the first part is dedicated to a particular reading of the notion of diagrammatics in science and creative work. Richard Feynman's unorthodox markings on a classroom blackboard, which he employed to help render sensible the mechanics of the quantum phenomenon, are here used as a catalyst to explore visualizing links between art and science. The artist Francis Bacon is used as a counterpoint to show the emergence of a tripartite language of diagrammatic-creative-intuitive in twentieth-century painting. The main focus here is Feynman's diagrammatically enhanced lectures and exploring various theoretical understandings of quantum mechanics that reveal new possibilities of reality through the making of graphic marks (1982). These different realities are informed by Gilles Deleuze's writing on the character and utility of diagrams (2005). Here, the diagram is given a new identity by the very nature of the immateriality of its quantum subject matter. The probability of a multiplicity of 'becomings' emerging from graphic vectors are explored and employed to reinforce the syncretic relationship between quantum theory and creativity.

Chapter 3: Spin – is devoted to the phenomenon of quantum spin in subatomic particles. Quantum spin is the phenomenon that makes quantum computing possible. This chapter explores quantum computing's predicted ability to rival and surpass human intelligence, which is at a critical cultural point to challenge and stimulate thought. The artistic practice-led research in this area unpacks and constructs poetic links with the potential for such things to exist beyond what is currently known or understood. Spin becomes a fundamental material connection to the preservation/archaeology and processing/recovering of data. The data is atomic and is synonymous with data processes of our own human atomic cognition:

atomic spin echoes the human brains and the quantum processing of information. The artwork of Paul Thomas and Kevin Raxworthy is explored to find the synchronistic relationship between quantum data and consciousness.

Chapter 4: The Graphene Moment – constitutes a case study of carbon graphite as an example of a material and medium that seems to never cease in offering up new material agencies. From the initial discovery of graphite deposits under a tree at Borrowdale in the United Kingdom, to the role of graphene sheets in new imaging technologies, carbon graphite has facilitated, through the technology that it has permitted, a feedback loop between material agency and successively deeper scientific inquiry. The art historical importance of graphite will be established in detail, as will the possibilities presented by the ongoing and radical reinvigoration of the medium in the quantum age. The agency-discovery feedback of materials is framed in this chapter as the possibility of material consciousness and object ontology. This chapter will explore artists working with graphite from Georges-Pierre Seurat to Joyce Hinterding, Anish Kapoor, Cornelia Parker, James Gimzewski, Victoria Vesna and Frederik de Wilde. Graphite may now serve not only to record and map thoughts, but also inscribe a range of non-semiotic utilities, such as its aptitude as a conductor of electromagnetic currents. The study of graphite is seeing the performative agency of matter, driving a reconnection between consciousness and (supposedly inert) materiality.

Chapter 5: Cloud – explores uncertainty as metaphorical distribution of 'atmospheric' storage and retrieval of information in distributed cloud-based networks. It explores the cloud as a central feature of uncertainty that would eventually lead to uncertainty being legitimized in 1927 by quantum mechanics.

The data cloud[9] exists everywhere and yet it is nowhere in particular. Like breathing in (downloading) and breathing out (uploading), the aesthetics of the data cloud are both organic and swarm-like. As with the protocols of bit torrent files, the cloud has effected the erasure of digital boundaries into boundary-less states of informational assembly. This final chapter explores the precept of atmospherics as one of both art historical and philosophical interest and agency. 'Cloud' draws inspiration from the work of J. M. W. Turner whose paintings seem to coalesce the distinct elements of composition into a swollen and undifferentiated atmospheric whole. In this chapter I aim to frame the uncertainty that is inherent within the analogous physical cloud to see it as a problem of perspectivalism that could not be absorbed within the objectification of space. I explore the cloud's atmospheric quantum qualities visualized in Turner's paintings and the work of Semiconductor as visualizations and philosophical intuitions of phenomenon.

All we know for certain is that both art and science have probability and uncertainty at their core. These seemingly disparate worlds of qualitative and quantitative research have been tenuously entangled for decades. This is due to artists continually asking complex questions that align with a scientific fascination for new discoveries and the increasing appreciation that creativity and subjectivity inform and parallel their objective processes and knowledge systems.

The fragments clustered around the themes in this book reside in the atmospherics of consciousness through the convergence of creativity and scientific inquiry. This book is a journey guided by the Epicurean swerve and the miniscule shift in direction that needs to occur to then bring the world into existence. The position of culture is redefined by uncertainty, as it seeks to untie itself from the predictability of measurable, deterministic, quantifiable science. The ability to speculate on the various hypotheses of humanity came directly from the unavoidable spin, the inevitable swerve of atomic forces, triumphed by the uncertainty principle. I describe the synchronistic events that take place by demonstrating invisible concepts within and surrounding quantum phenomena, and examining how the various forces attract and shuffle, independently and singularly, yet simultaneously together.

Notes

1 Speculative realism is a heading under which philosophical groups share a common concern where they all appear to disagree with the Lacanian concept of human-defined boundaries. Each group of speculative realists, speculative materialist and object-oriented ontologists have basic philosophical differences of approach to human boundaries. Object-oriented ontology is a philosophical movement that does not privilege humans over non-human objects, or thought over dead matter. This concept is a recurring theme of the book, thinking between human and non-human.

2 The difference between quantum physics and quantum mechanics is related to history when Niels Bohr proposed the concept of quantum physics in 1913. The name was changed by Heisenberg and Schrödinger to quantum mechanics. So, while Bohr's explanation became based around the title of quantum theory, all other burgeoning subheadings fitted under the umbrella title of quantum physics. Quantum mechanics is seen in this book as being fundamentally the mathematical portrayal of the position, motion and interaction of subatomic particles. The field incorporates concepts of quantization of energy, wave–particle duality, the uncertainty principle and the correspondence principle.

3 The word quantum (plural: quanta) came from the Latin meaning 'how much' or 'how great'. Quantum is a discrete quantity of energy proportional in magnitude to the frequency of the radiation it represents. It is seen as being the minimum amount of any physical entity involved in an interaction. For example, a photon is a single quantum of light.

4 My memories of hearing as a young man that you could get all of human knowledge on the head of pin was written by Richard Feynman, who was unknown to me at that time as a leading proponent of quantum mechanics (Feynman 1960).

5 A 'becoming' in this book is seen as something actualizing over time, linked to Bergson's concept who states: 'Matter or mind, reality has appeared to us as a perpetual becoming' (Bergson 1911: 287).

6 Thomas Young's experiments to measure light waves created a double slit experiment that demonstrated for the first time an interference pattern (1801) (Jaeger 2009). In relationship to the ongoing controversy of the double slit experiment Feynman said in the first chapter,

volume three of his *Lectures on Physics*: 'We choose to examine a phenomenon which is impossible, absolutely impossible, to explain in any classical way, and which has in it the heart of quantum mechanics. In reality, it contains the only mystery. We cannot make the mystery go away by "explaining" how it works. We will just tell you how it works. In telling you how it works we will have told you about the basic peculiarities of all quantum mechanics' (Feynman 1963).

7 Taken from a transcript of a video from Feynman's lecture at Cornell University entitled *Probability and Uncertainty: The Quantum Mechanical View of Nature | The Character of Physical Law* (Feynman 1964).

8 Epicureanism is based upon the teachings of the ancient Greek philosopher Epicurus, an atomic materialist who built his ideas on the thinking of philosophers such as Democritus. The Epicureans developed a philosophy that was considered hedonistic, but was actually based on pleasure and a simple life. The perceptions and critical inquiry of the world around them were fundamental to the great insights in to the material world. As Michael James Bennett explains, 'Epicurus contends that there is a kind of continuity between perception and conception. Having established that atoms are the principles of all things, Epicurus' *Letter to Herodotus* describes the ceaseless character of atomic movement' (M. J. Bennett 2013: 132).

9 The data cloud and cloud computing are terms coined in 1996 at the offices of Compaq Computing, which were brought back into focus by Eric Schmidt, CEO of Google, in 2006. The term is generally used in referring to accessing and using data remotely (Paul D. Miller 2014: xx).

Chapter 1

The Swerve

Agency of the swerve

To propel and maintain us moving forward into concepts of agency in the Epicurean swerve, we might employ the concepts of 'vitalism'[1] or 'entelechy'[2] that holds that there is an inherent life within various natural materials, which allows for and appreciates the cognitive comprehension of matter as a performative agent.[3] The performative 'spin' (to appropriate the quantum analogy into a creative one) is at the heart of the matter here – along with vibration, rhythm and material patterns, it is as fundamental to the physical world (indeed it is the fundament per se) as it is to creativity and consciousness. Like Schrödinger's cat, we are compelled to understand materiality as being both alive and dead simultaneously. A non-human object like a cup takes on a humanness by virtue of its function; it becomes a semi-alive companion that exists in the space where it is not affirmed.

It follows that all possible futures emerge out of an undetermined past. This is something we could know implicitly (self-evidently) if we were possessed by a higher dimensional consciousness. Within our narrow vision, all we can hope to witness is the flickering of possible futures behind the glare of the one we create. In this sense, modelling and rationalizing the world is a kind of foresight or divination, whereby our potential futures emerge from the unknowable yet-to-be and take first cautious and underdeveloped steps in the present.

The prolific and popular role of scientific storytelling (popular science) in today's world means that we are more aware of the importance of the vibratory character of existence. Yet, this coalescence of creativity and science (for better or worse) has a long lineage. Based on the work preceding the Solvay conference, the Italian Futurists sought to give visual form to new concepts of matter emerging at the time and concerning the newly discovered quantum world. This world is elucidated in the writings of Henri Bergson and the technical images of Étienne-Jules Marey[4] by drawing together a conception in seeing the totality of space and the object world as an exemplar of contributing fields of energy.

We can draw further relationships between Marey's chronophotography of movement and the Atomic Force Microscope's visualization of atoms. The moments in time when the invisible worlds become visible seem to spark universal assimilation of an expression of movement by common interest. To either extol or challenge the mechanization of visualizing the invisible is the question here. The late nineteenth-century concept of 'ether' as an invisible, nonmaterial and infinite space filling the void between things becomes a tangible influence in Marey's photographs, revealing a confluence between scientific storytelling and artistic world-modelling. Marey's photographs utilized the fabric black velvet as defining an analogous and relativistic pictorial expression of blackness to emphasize the space between. The movement of the subject was captured by the camera in front of the low light emitted from the blackness of the velvet. What was then seen in the photograph were traces of the subject coalescing within the fabric to form a physical representational space of ether.[5]

The Italian Futurists showed us in the early 1900s the initial stages of how concepts being explored within physics were being investigated through art simultaneously. The courage to visualize the invisible, but foundational, material world is seen in the work of Futurist artists who drew on Marey's photographs, as well as Bergson's theories.[6] Giacomo Balla, Umberto Boccioni, Gino Severini and Carlo Carrà all attempted to reveal in their work the imperceptible forces of the atomic world. They sought to explore and represent (albeit in what we now understand as traditional mediums) new ideas about time, space and motion as being inclusive of one another; perceptible but not visible. Boccioni was making the most sensational contribution to this project in his attempts to articulate a new visual language of matter.[7] The state of making such profound works was not solely driven by physics but was part of a total experience of critiquing a mechanistic age. Fearful of the old and intoxicated with the new, the Futurists were testing out creative possibilities of visual expression against a backdrop of revolutionary scientific advancement.

Measurement and control

Measurement is continually challenged in the speculative nature of the quantum universe. Eight years after Einstein's theory of relativity, and fourteen years before Werner Heisenberg articulated and formulated his quantum uncertainty principle, Duchamp poetically demonstrated the problem of measurements in his work *3 Standard Stoppages* (1913–14). The contestable nature of measurement is clearly defined in Duchamp's artwork, which demonstrates the artist's intuitive vision and disregard for the classical picture of the world. In this experiment, Duchamp took three pieces of yarn, each a metre long, and dropped them onto a length of Prussian blue canvas. Each time the yarn was dropped, it formed a different wave; each piece of string expressed a new and random version of the length of a metre. These different random waves were marked up and cut out of wood to form three correlate templates, which were each resultantly different lengths. The metre had been a standard measurement since its introduction in France in 1793; Duchamp successfully interrogated the absolute certainty of classically defined units of measurements. *3 Standard Stoppages* was a reflection of its time as a critique of measurement.

Duchamp's experiment literally touches on the influence of gravity and time and impacts Newton's laws of motion. The certainty of the Newtonian world was unravelling by the evolving problem of measurement at the atomic level. Physicists during Duchamp's time were questioning the legacy of the classical world. The instruments of measurement created in *3 Standard Stoppages* can be seen as representational of the uncertainty principle and superposition,[8] with the metre outcome being in multiple states at the same time.

In an interview with Katherine Kuh, when asked which of his works he finds most important, Duchamp stated:

> As far as date is concerned I'd say the *Three Stoppages* [*3 stoppages étalon*] of 1913. That was really when I tapped the mainspring of my future. In itself it was not an important work of art, but for me it opened the way – the way to escape from those traditional methods of expression long associated with art. I didn't realize at that time exactly what I had stumbled on. When you tap something you don't always recognize the sound. That's apt to come later. For me the *Three Stoppages* was a first gesture liberating me from the past.
>
> (Kuh 1962: 81)

Duchamp's artistic provocation sought to liberate artists so that they might explore those new territories that were being opened up by the sciences. The quantum and the atomic, Duchamp realized, needed to be co-opted and contextualized in a comparative study with what artists had already historically accomplished in the confrontation of measurement. This emerging confrontational trend can be seen as one that expresses at its core both scientifically inspired ontology (being) and epistemology (knowing). In many respects, Duchamp and those kindred artists were making Schrödinger's cat – thought experiment – pre-actively into an artwork.

To quote Lewis Feuer:

> The arts were probably the most readable barometer of the revolt against determinism during the war and post war period. This was especially evident in the turbulent emotional assertions of the Dadaist movement. Here, almost as if in a pure laboratory culture, was the emotional nucleus of the rebellion against causality, order, and logic.
>
> (Feuer 1974: 183–84)

The act of drawing out and critiquing the world through mark-making is an experimental methodology for understanding our existence. Duchamp's work is insightful because it challenges orthodox classical models and in doing so reveals the probabilistic nature of quantum. Without knowing it, Duchamp was tapping into the fundamental uncertainty borne of the quantum wave/particle duality in his reproach and critique of inherited standards of measurement. A new non-deterministic atomic behaviour creates a special kind of unpredictability for the neutrons, photons and electrons to be understood as being part of the unmeasurable quagmire at the foundation of nature. Marcel Duchamp captured the twilight of the classical idols succinctly, using yarn to create diagrammatical linear elements to be reused as an uncertainty wave for reference in subsequent paintings.[9]

There is never just one sole event that leads to the discovery of a particular affect. The development of chance or probability challenged the quantifiable world to realize that simultaneous occurrences of a particular problem are not only possible but also necessary. The artistic movements of Expressionism, Cubism, Futurism, Constructivism, Dada and Surrealism all included artists who were looking at questions that pertained to physics, yet did so often without recognizing it; certainly, physics and chemistry were not the sole focus of their intentions. Picasso's interest in primitive masks, for example, was not simply the result of oriental imaginings; his interest in them was, in part, fuelled by the ongoing race to represent four-dimensional space (a fact that is well documented by Arthur I. Miller[10] and Linda Dalrymple Henderson[11]).

The Solvay rupture

The Solvay International Conference was a highly significant series, initially instigated by Belgian industrialist Ernest Solvay in 1912. The most famous of the conference series (that last took place in 2014) was the 'Fifth Solvay International Conference on Electrons and Photons' (October 1927) where the world's most notable physicists discussed quantum mechanics. Here, the legendary Bohr–Einstein debate began on the probabilistic nature of quantum mechanics. The conference was where Einstein, dissatisfied with Heisenberg's uncertainty principle, infamously made reference to suggest that 'God did not play dice', to which Bohr supposedly retorted 'stop telling God what to do'.[12] Whether this story is true or not, it points out the immense differences that were generated between the classical and quantum worlds. This particular conference was the culmination of a shift in thinking that would have major causal effect across the world.

Located in Brussels, this conference was where the most notable physicists, including Niels Bohr, Paul Dirac, Albert Einstein, Louis de Broglie, Werner Heisenberg, Max Planck and Erwin Schrödinger, were challenging prevailing givens within physics and chemistry.

Werner Heisenberg's paper 'On the Perceptual Content of Quantum Theoretical Kinematics and Mechanics' (1927) articulated and defined what we now know as the uncertainty or indeterminacy principle; a principle that holds that if you can measure the velocity of an atomic particle, you cannot measure its position in space on account of the physical fact of the wave–particle duality of all matter. The history of the events that followed the 1927 Solvay conference is complex but might be summarized or threaded together by what we would now call a deconstructivist trend – although it was a trend not to combat the validity of scientific processes, but to combat the fact of scientific certainty. In other words, the 1927 Solvay conference (Fig. 1) seeded a sense of doubt and stressed the importance of the latent uncertainty that undermines our ability to measure anything with absolute objectivity.

This uncertainty was the basis for the famous Copenhagen Interpretation,[13] created to define a means by which the ambiguity that surrounded the findings of Heisenberg and Bohr could be contextualized. The Interpretation made a role for probabilities in measuring an event that does not have definite properties, which meant that the experiment was always reduced to one probability for a classical outcome. It was this decision of the Copenhagen Interpretation for the uncertainty of the wave-particle duality in physics to exist as though two unmeasurable sides of the same coin. This created the fertile grounds for uncertainty to exist as the essential core of quantum theory.

Albert Einstein, Boris Podolski and Nathan Rosen (EPR) developed a theory in 1935 which disputed the Heisenberg theory of uncertainty. Later, in 1964, at CERN in Geneva, John Bell developed a paper that hypothesized a way to prove or disprove quantum mechanics, creating a fundamental construct that enabled physicists to finally come to terms with certainty/uncertainty as a physical fact of the universe. Bell's hypothesis was to capture reality in the act of happening.[14] Einstein famously called quantum mechanics 'spooky action at a distance'.[15] 1927 thus serves as an important historical junction in the

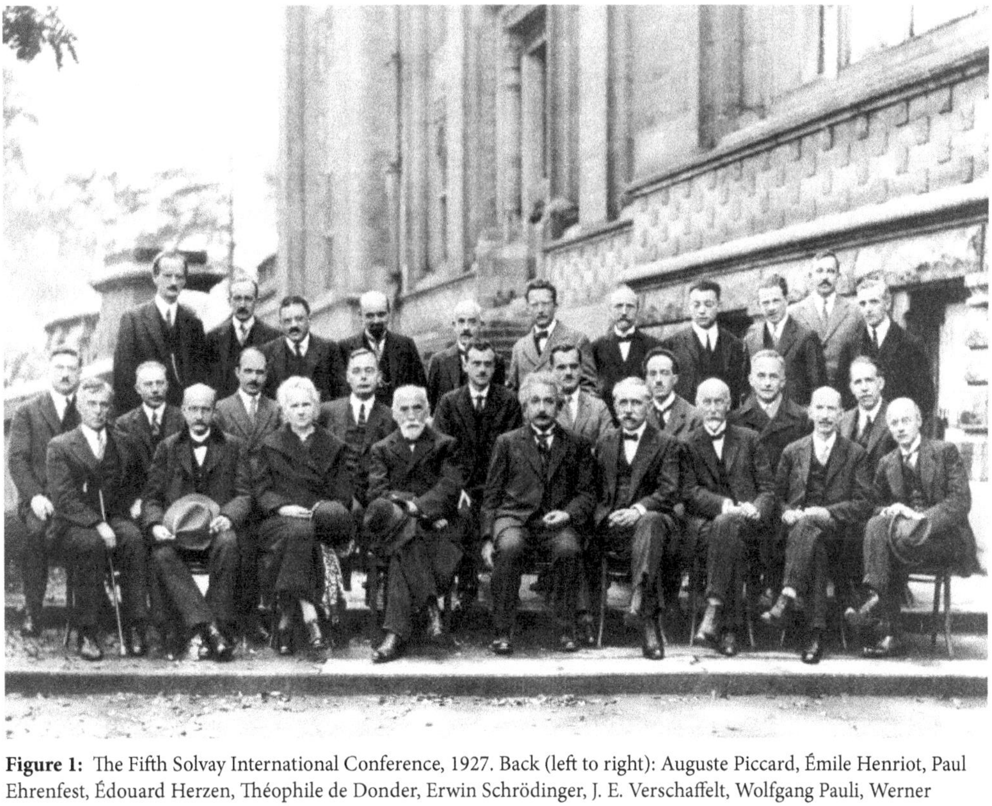

Figure 1: The Fifth Solvay International Conference, 1927. Back (left to right): Auguste Piccard, Émile Henriot, Paul Ehrenfest, Édouard Herzen, Théophile de Donder, Erwin Schrödinger, J. E. Verschaffelt, Wolfgang Pauli, Werner Heisenberg, Ralph Fowler, Léon Brillouin. Middle: Peter Debye, Martin Knudsen, William Lawrence Bragg, Hendrik Anthony Kramers, Paul Dirac, Arthur Compton, Louis de Broglie, Max Born, Niels Bohr. Front: Irving Langmuir, Max Planck, Marie Curie, Hendrik Lorentz, Albert Einstein, Paul Langevin, Charles-Eugène Guye, CTR Wilson, Owen Richardson. ETH-Bibliothek Zürich, Bildarchiv. Photographer: Benjamin Couprie, 1927.

development of our understanding of the physical universe. Since then, within scientific and creative fields alike, there has been space for the impossible, unpredictable, unplanned and random to play an essential role as part of intuition and inquiry aimed at making sense of what it is to be in and of this world.

The path of 'uncertainty' precipitated in 1927 was nevertheless met with resistance from the orthodox scientific community. 1927 saw institutional conflict emerge between the two philosophical camps of inquiry – between certainty and uncertainty. This conflict persisted even after Bell's theory of 1964. In 1982, when the technology was available, Alain Aspect, Jean Dalibard and Gérard Roger created the actual experiment inspired by Bell to see if the world we live in is probabilistic (speculative) or classical (local realism), where the world has definite things and only interacts with neighbouring states. Aspect's paper 'Experimental test of Bell's inequalities using time-varying analyzers' (1982) shows, by using two entangled electrons, that the predictions of quantum mechanics are not compatible with the classical world (Aspect et al. 1982). Barad articulated the problem as follows:

> On the face of it, these questions seem vacuous, but there may be more here than meets the eye. Consider, first, setting up the condition for the experiment: we begin with a vacuum. Now, if a vacuum is the absence of everything, of all matter, how can we be sure that we have nothing at hand? We'll need to do a measurement to confirm this. We could shine a flashlight on the vacuum, or use some other probe, but that would introduce at least one photon (quantum of light) onto the scene, thereby destroying the very conditions we seek. Like turning up the light to see the darkness, this situation is reminiscent of the mutually exclusive conditions of im/possibility that are at issue in Niels Bohr's interpretation of quantum physics.
>
> (Barad 2012c: 547)

Barad's notion of light here resonates with Martin Creed's installation *Work No. 227: The Lights Going On and Off* (2000–01) (Fig. 2), awarded the prestigious Turner Prize in 2001. Both of these 'descriptions' reflect the fundamental problem of the measurement of our spatial surrounds by the admittance of light to darkness. Here, we see an intuitive link between quantum theory and conceptual art. In his installation work, Creed set the lights on and off in five-second intervals. He explains that:

> It activates the whole of the space it occupies without anything physically being added and I like that because in a way it's a really big work with nothing being there [...] My work is about fifty per cent what I make of it and fifty per cent what people make of it. Meanings are made in people's heads. I can't control them.
>
> (Reynolds 2001)

In this context, we can see the parallels that are drawn between Barad's insights with the problems of measurement within the world of quantum phenomena. The statement from

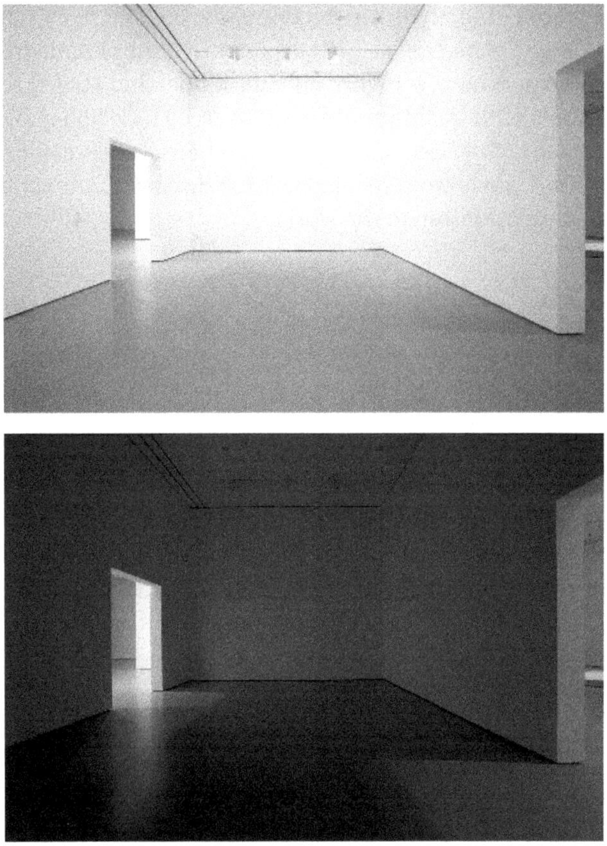

Figure 2: Martin Creed, *Work No. 227: The Lights Going On and Off*, 2000–01. Installation. Courtesy of Tate Photography © Martin Creed.

Creed also posits a view that is reflected in the ideology of Quantum Bayesianism (QBism[16]) in how one's sense of the world coming into existence is based exactly on the percentage that you make it (QBism will be discussed later in the chapter). Creed's work reflects the uncertainty of existence and follows on from a lineage of inquiry concerning things and how we interact with them. Things, all things, come into being by light; when nothing is placed in the room we only see the light coming that brings the space into existence. The light waves that bring all things into being form the basis of a perceptual becoming. At the quantum level, Creed's work reveals the impact of measurement in the construction of our understanding of the material world.

If we shine a light equivalent to one photon, we influence the quantum state of things and may therefore inadvertently create a classical state. A given thing existing in a dark space would have the same problem, but in relation to our haptic interactions. We touch, and by that very nature of touching, we bring the thing into a classical existence. Creed's *Work No. 227: The Lights Going On and Off* (Fig. 2) explores in a simple construction what might be seen as a thought experiment, a highly difficult model and structure of the interplays between space and time, with quantum phenomena at its indecipherable core. The legacy of Duchamp's *3 Standard Stoppages* is alive and well here, which can be seen as having a direct relationship with Creed's work insofar as it too deals with the fundamental problem of measurement in our relationship to the world.

Capturing reality

How can art seek to explain, through methods and mediums, a new voice, a new language that creates an empirical awareness of its role in the development and influence of thinking about the nature of our world? Gavin Parkinson in his article 'Surrealism and quantum mechanics: Dispersal and fragmentation in art, life, and physics' situates this argument historically and culturally with reference to the surrealist artists of the 1930s. Parkinson establishes an argument in which he highlights the complexity for artists to develop a language for the 'new reality' being constantly revealed by the physicists. The relocation of surrealism in a discursive field that includes quantum physics opens up for its artists' shift to a semi-abstract language (2004: 557).

Parkinson particularly focuses on the art and context of Wolfgang Paalen and Roberto Matta:

> The broad-ranging exchange that took place within the avant-garde and academia in 1930s France on the new challenges laid to traditional forms of rationalism by modern physics, psychology, philosophy, and art, is one of those awaiting a reception in the scholarship.
>
> (Parkinson 2004: 557)

In the painting *The Vertigo of Eros* (1944) by Matta, we can see an attempt to draw from modern physics and psychology, visualizing the effect of these two different disciplines. The capability of the artist is severely tested when confronted with attempting to represent the inner and outer worlds that are both in a state of uncertainty and unpredictability. The artist internalizes the totality of the information to translate it into a new visual expression that attempts to summarize the world. Freud's study *The Ego and the Id*, first published in 1923, is situated near the time of 1927 and Heisenberg's uncertainty principle.[17] These two major historical shifts question how we internalize the world; Matta's work offers a visualization of these questions.

Parkinson's thinking here reveals a link between the illuminations gained from probing and constructing realities at the same moment that science is exploring its constructed realities, drawing out synergies in art, science and technology. The underlying connection between these pursuits and examples is the desired outcome – the desire to capture reality as it happens or in the act of happening. This task will be further explored in two creative examples – the work of Sir William Coldstream and Alberto Giacometti.

Coldstream's creative methodology was formative in my personal experience of learning how to draw as a student, the implications of which are all the more prevalent to me now. Coldstream developed a process to measure every relevant point when examining a still-life installation, and while mapping and transferring that multi-nodal image to the canvas or paper (Fig. 3). This methodology (which in one instance took him over two years) remains visible in the finished work, as each sitting demanded a new examination of the reality being confronted. The artistic practice developed ultimately reveals the fundamental insecurity of analysis, beset by human fallibility, as the errors in its processes are identifiable as visible graphical mistakes. The measurements concerned with how long or how diligently the artist is working are, in this case, subject to human imperfections when it comes to representing what we see on a two-dimensional surface. In constantly making marks to capture reality as it is happening, the painting ultimately fails itself, as that reality keeps evolving. This impossibility of reality as a finality in Coldstream's practice resonates with what is at stake in quantum mechanics and its uncertainty, and relates (in terms of outcome) to Bell's aforementioned desire to prove or disprove quantum theory in reference to measurement. To reiterate this problem, we return to Barad:

> If the measurement intra-action plays a constitutive role in what is measured, then it matters how something is explored. In fact, this is born out empirically in experiments with matter (and energy): when electrons (or light) are measured using one kind of apparatus, they are waves; if they are measured in a complementary way, they are particles.
>
> (Barad 2012c: 647)

In this context, we are looking at different ways of observing the world and bringing it into formal or conceivable existence. The author Wojciech Hubert Żurek, in his paper 'Quantum Darwinism', articulated quantum processes as a physics of evolution concerning the relationship between a system and environment:

> Upon measurement, a general preexisting quantum state is erased – it 'collapses' into an eigenstate of the measured observable. How is it then possible that objects we deal with can be safely observed, even though their basic building blocks are quantum?
>
> (Żurek 2009: 1)

Observation is the act of bringing the world into existence, albeit as a local and localized classical reality. Realism, as an ideological pursuit, is therefore universally impossible as long as the observer is localized. The classical world and the quantum world cannot exist together because where one exists as physical fact, the other is underwritten by the

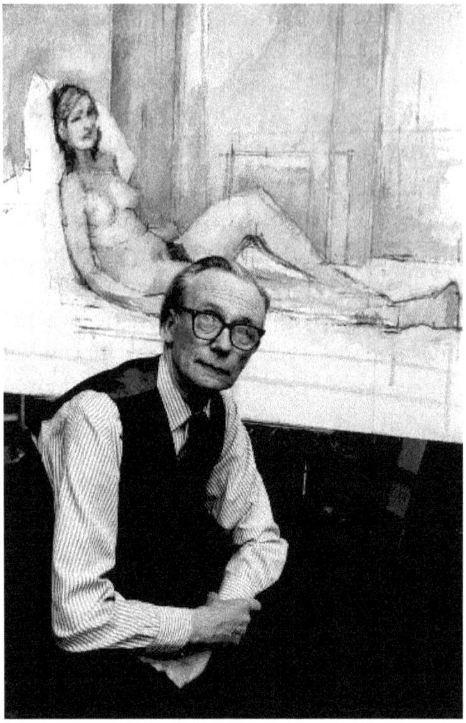

Figure 3: Studio photograph of Sir William Coldstream in front of *Reclining Nude*, 1975. Photograph by Harry Diamond.

physical fact of uncertainty. The artist-as-observer – a unique breed of observers – needs to be qualified in terms of Epicurean optimism qua the possibility of an empirical understanding of the world. Here, the artist is aware of himself/herself as an instrument that comprehends the world as being more than what was first thought. The artist-as-observer creates the world and adds to it in parts. It is these parts that are classical fragments that are assembling, disassembling and reassembling together, becoming a whole to the observer. The fragments are frequencies that penetrate and permeate our bodies, interpreted as an ongoing stream of data reflecting current actualities. Every interpretative shift changes the world-as-perception in some way. Thus, observational and creative inquiry into the world serves to expand and explode consciousness with new meanings and questions that challenge the human understanding of the nature of the physical universe. Coldstream's influence in the area of observational and creative inquiry is found in the work of one of his ex-students, Ewin Uglow, who studied under Coldstream at Camberwell School of Art. The democratizing of space that Coldstream used to assist with the concepts of measurement can be seen in Uglow's studio (Fig. 4). Every item forms an indexical relationship with the absent figure marked out on the ground to be

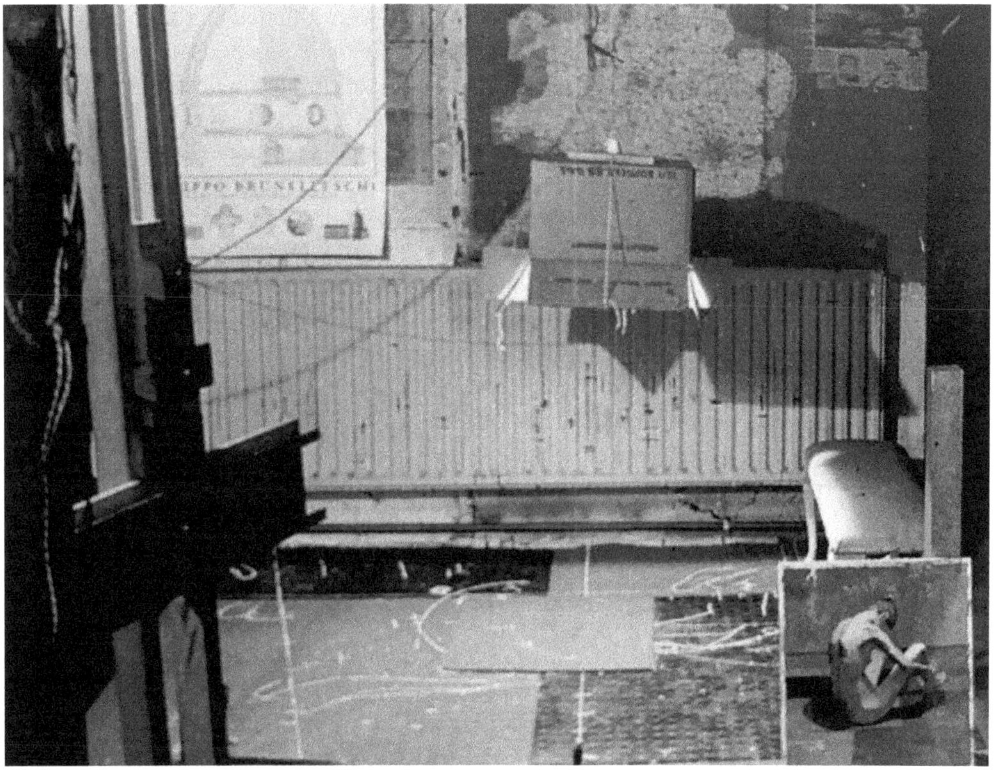

Figure 4: Photograph of Ewen Uglo's studio. © Browse & Darby Ltd.

painted. The radiator, gridded map, tiled floor and horizontal links of the skirting board and the seat are all waiting for the life model to resume a position within the boundaries of the taped markings on the floor.

The tools of measurement such as the tape and the plumb line assist in trying to draw reality. The inability of the artist to capture reality in the act of happening uses the very probabilities inherent within quantum physics to continually reassess what the world might be.

Coldstream had a major influence in British art schools, where his drawing technique challenged the total view of perspectivalism. His methodology revealed the interconnectedness of all determinist positions that define the subject of an image and its pictorial boundary. The recognition of any proportion, negative space, vertical or horizontal structures that were incorrectly drawn led to a string of changes that always ended up referencing the original mark. The marks had to be relevant to what was seen and they were made indexical to the object of the artist-as-observer's analysis. Coldstream's practice emphasizes that analytic focus on one particular area of a given scene means the artist is no longer drawing the whole picture – therefore, the artist is forever trapped in a process of collapsing multiple images into a synthetic unified one.

The measurement methods endorsed by Coldstream were situated as part of a broader and emergent field of anti-perspectival consciousness. Coldstream's practice made it clear that any unique set of marks drawn directly from observation have a fleeting and limited relationship with reality. A given mark in itself is laden with all possible meanings that relate it to its position and place within the creative matrix of canvas or paper. The artist's mark creates the probability that this is the correct position; the second mark needs to substantiate the position of the first and so on. These highly complex (classical) processes of drawing marks that deal with a multitude of different material agendas, atmospheric conditions and spatial optics creates more problems than solutions. The complexity of problems generated multiplies when attempting to capture the classical world as it is happening, with each consecutive mark attempting to bring about a becoming. It is the classical presumptions of a stable and certain physical world that Coldstream continually attempted to build and dismantle in his work. Coldstream's process brings into account all possible conditions of the object (of the image) and the infinite data of the surroundings that the artist is drawing from – the interplays of gravity, atmospherics, material quality – resonate within the image; thus, the search for reality through the drawing is impossible to rectify in finite space. The space that surrounds all objects is itself brimming with information and murmuration.

The observer, in trying to measure and contest the classical world, will only discover that their observations are always and necessarily inadequate. Each work of art is a theoretical supposition of the world that is observed, or as it looks like when brought into the pictorial plane. Artworks that reveal the operations of their own constructedness then lend themselves to the fundamental reality of uncertainty that they expose and embrace ideologically.

In 1933 Coldstream was part of a short-lived British experimental art movement called 'Objective Abstraction', whose main premise was based on a version of what we now call 'object-oriented ontologies' (OOO). The object and its immediate spatial environment are conspiratorial in their dependence on each other for their existence. The concept to draw out all the possibilities of the object and express them abstractly is another attempt to find a language that inherently describes the uncertainty of the world. The desire to explore this area of investigation reflects how other art movements were intuitively internalizing the growing understanding that was taking place in the world of physics. Objective Abstraction demonstrated a desire to more fully understand the observer's democratic gaze in relationship to the world as something more than a fleeting glimpse. Coldstream's ontology of uncertainty was derived from watching the world and accepting that it evolved into more and more complex and less certain significance, instead of, as classical thinkers proposed, emerging slowly but surely as certainty and clarity in representation.

The movement only appeared to last for three years, with a major exhibition called *Objective Abstraction* taking place at the Zwemmer Gallery in 1934. The complexity of developing a mode of visualization that was equivalent to an elegant equation was a challenge. Concerning the Objective Abstraction of Coldstream and his peers, John Wilson writes that they 'shared with Giacometti the oppressive conviction that what he was trying to do was impossible and yet I think he was not too displeased with the painting itself' (Wilson 1999: 16). The collapse of the classical divide between meaning (art/creativity) and measurement (science/observation) is central. We can see this specifically in the drawings of Giacometti, an artist attempting to capture in his two-dimensional work reality as it happens. The marks in Giacometti's drawings, like Coldstream's, are subject to human imperfections when it comes to representing what we see on a two-dimensional surface. In constantly making marks to capture reality as it happens, the drawings fail by the very act of measuring that reality. This process draws parallels in the same way that quantum mechanics indexically references measurement.

Coldstream, who was then head of the Slade School of Fine Art in London, asked David Sylvester to give regular lectures. Sylvester organized the Giacometti exhibition at the Arts Council Galleries in 1955, feeling that there was a strong connection between the work that Coldstream was doing and the Futurist experiments of Giacometti, thus enabling a meeting to take place between the artists and their work (Laughton 2009: 69). The connections that Sylvester perceived between the two artists, which he documented in his catalogue essay for the show, were qualified in terms of a common interest in cultivating 'something more than representation' (Laughton 2009: 69). Sylvester elaborates on this resonance in greater detail:

> The most striking thing about the paintings, and to a lesser extent the drawings, is the density of their space. The atmosphere is not transparent: it is as visible as the solid forms it surrounds, almost as tangible. Furthermore, it is uncertain where solid form ends and space begins. Between mass and space there is a kind of interpenetration.
>
> (Sylvester 2001: 54)

For his part, Giacometti had a major influence on drawing techniques. His experimental methodology produced images that suggested a conglomerate and multi-perspectival view, wherein each individual observation played a part in attempting to define the position of every other part. The recognition of any proportion that was incorrect led to a string of changes that always questioned the original point of reference. The marks had to be relevant to the fugitive nature of what was being seen and they were made in an indexical fashion relative to what was being questioned and understood by the artist. Giacometti's marks led to more speculative and ontologically pertinent questions, tracking spatial norms only to dispense with them.

To work in the methodological process of Coldstream and Giacometti (Fig. 5), one cannot fixate their gaze on a particular area or aspect of an illustration or image of the world – the object that is the focus of creative inquiry. Such ontological drawings are no longer problem finding but problem solving. The artist creates blinkers that direct his/her perception. As with diagrammatical imagery, the uncertainty of the world comes into being in object ontological imaging. The space that surrounds all the objects – the ether, dark matter, the void – is recognized in this breed of imaging as the very substance of that which is being imaged. Each object and the ether that surrounds it is seen, and brought to the mind through association of memory that interprets them; they are thus, to quote Bergson, 'drawn to the surface by the perceptions which resemble them' (1912: 10).

We recognize a symbiosis of what is constructed in the mind, between what we perceive and what we know. Each mark bears an ontological understanding of what is observed (hylomorphism) and understood. In other words, the ontological being – the object – is recognized and simultaneously its existence is challenged in context to the whole it is

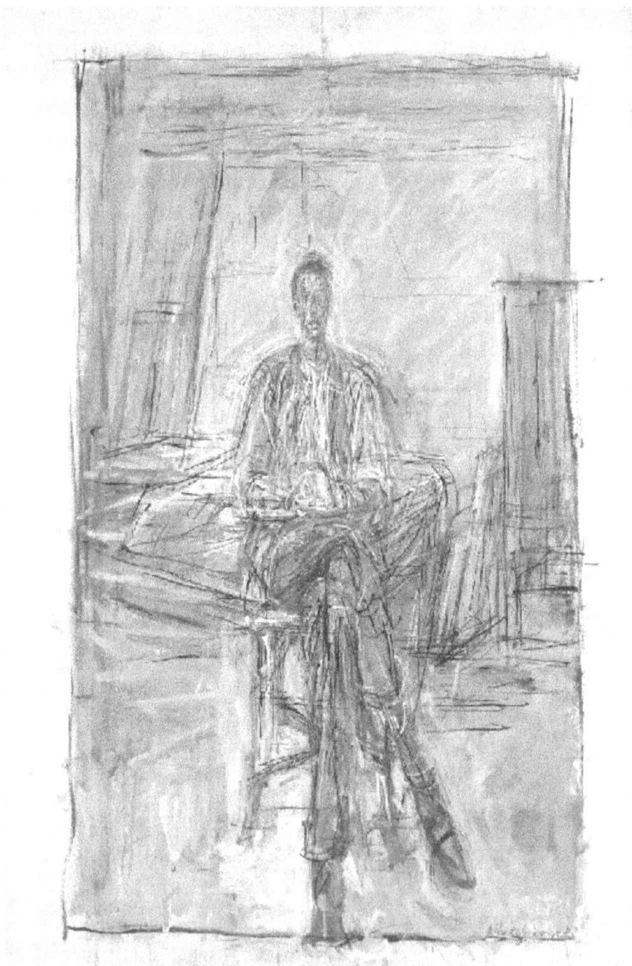

Figure 5: Alberto Giacometti, *Seated Man*, 1949. Oil on canvas.
© Fondation Alberto et Annette Giacometti/Licensed by Viscopy, 2018.

experienced in. The mark then is a process of updating an understanding of uncertainty of that being, through the fundamental yet interminable quest of perception to respond and bring the world into existence.

In trying to measure and contest the classical world, the observer-as-artist discovers that their reflections are always inadequate, as the world operates as a coalition of uncertainties. Scientific inquiry, in the same way, attempts to approximate and quantify the phenomena of the void, and, in so doing, deny the void's existence. This brings us back to the reference from Barad 'Like turning up the light to see the darkness'. The denial is one where the object of the measurement is only challenged as an external element, not a being of the object. The objectification of the material world is significant in attempting consistency of experience that should not be polluted with associated memories.

Contemporary art is drawn from questioning how the world is; to explore our experiential relationship about how things come into existence; to explore all the different subtle variations that come from being human and being of this world, this universe. How artists engage with the world through their senses constitutes a form of measurement. By the very act of observing the world, artists are bringing the world into existence. This world is personal, that is, not a separate entity to and from being human, but a co-constituted framework of mutual simulacra. To enter into this process of continually challenging what we experience when we focus our attention to the question of interacting with the world, we indirectly become that part in the experiment that pollutes it. We become the vehicle that translates the world from chaos to try to define a form of certainty.

The unpredictable swerve

In his book *The Birth of Physics*, Michel Serres explains how the clinamen appears as a logical absurdity devoid of justifications; an instance in which its own cause – the cause of itself – is the same as the cause of all things. This ancient discourse of perception and measurement (physics) sees the clinamen as the cosmic tour de force. To comprehend the clinamen is to comprehend the universe's very existence as one predicated on a material disturbance; an atomic deviation. The clinamen is consequently manifest in all these sudden perturbations that come into existence in creative artistic production. The very flick of Jackson Pollock's brush released forces that are still being understood today, steeped with the history of the clinamen's cosmic agency cum the unpredictability of the atomic swerve.

The clinamen works across both the logical and the illogical comprehension of its own existence, finding its role in the space between metaphysics and physics. The connection between art and physics through the ancient model of atomism[18] becomes clear when we take various aspects of the atomic swerve (clinamen) as bringing the object-world reality into physical existence. For instance, John Bell's theorem is tested against the very act of the atomic swerve that creates perturbance, given that, as Lucretius suggests, 'atoms must a little swerve at times. But only the least, lest we should seem to feign' (Carus 1950: 55). Perception of this swerve that takes place between existing and feigning is the point where the artwork as an entity is made sensible and functional. The point where that which is being brought into the world is comprehended as (previously) an invisible element, which has been made visible or is being made visible, transcends the historical limits of representation as a fraudulent and futile self-deception to one that openly and transparently generates interaction with the void.

The void is essentially, in the Epicurean sense, the falling atoms as a consistent stream of physical matter. Here, aspects of the space are maintained as equal and opposite in a set of material interactions that maintains the status quo. The void is perpetual motion akin to the expansion of the universe; it is forever waiting for a shift in inclination. The swerve is the realization of this shift; a rupture in the constancy of the material flow of the changes in the universe. The haecceity of a single atom that changes its direction creates a convergence and instantiates a happening – all it takes is the primary swerve of one atom to create life. The change of direction can be modelled in the mundane example of street traffic. In this instance, the void is a stream of bodies walking down the same street at the same rate with no change of direction; they are creating a flow, yet they remain individual. Like these humans, Epicurus's atoms are systemically unified by the ongoing flow of traffic, imbued with the potential probability of disruption with even the smallest variance. Vortices, laminar, the clinamen or the swerve – these are minute deviation in directions. How do we represent this smallest movement of the atom? The most minimal of gestural marks is needed to enable the probability of a becoming to take place. This is the artist's concern: to witness the moments of synthesis in and with the world. The following quote from Charles S. Peirce's essay 'The doctrine of necessity examined' (1892) reveals synergies emerging in art and science at the time: 'Epicurus, in revising the atomic doctrine and repairing its defences, found himself obliged to suppose that atoms swerve from their courses by spontaneous chance; and thereby he conferred upon the theory life and entelechy' (Peirce 1892: 321).[19] The scientific concept of chance spontaneously links the arts and sciences together in their ongoing questioning of intangible forces.

The artist brings the world into existence through the on-going analysis of those constitutive particles that make up the potential of the thing being perceived in the first place. The mark, that is the infinitesimal swerve, is unrecognizable in its action but visibly confluent with what is being questioned, perceived and visualized. It is a becoming; a known without knowing why, a sudden rupture in the surface of the work that affects every other aspect (but not only) of the work. This rupture in the artwork is sensible in terms of Edward Lorenz's 'butterfly-effect' theorem whereby the slightest graphic ripple has the potential to cause a much larger cultural-cosmic upheaval.

We can model the tabula rasa of a blank white sheet of paper as an analogy to the void. Therein is nothingness, emptiness, and also infinite presence and possibility. The latent potential of the blank sheet of paper emulates a vacuum energy field. This 'plane of immanence' is not inert. The void is only tangible or interpreted where there is no perceivable disturbance on the surface, but still the surface oscillates, vibrates, and is both responsive and active at its atomic core. The artist disturbs the void by the act of making a mark that approximates some form of intervention into the cosmos: an attempt at measurement in which the mark collapses the probability of things and brings some given and static form into existence. The matter of the void is replete with its own potential for internal, invisible and yet latent perturbations, and maintains its own absolute atomic agency. Therefore, the problem with the mark in the void is that it assumes the collapse of the quantum world into the classical world we are living in. The imagination of the quantum world might then be understood by recognizing the mark not as a measurement by the artist, but as a simultaneous process (a fundamentally problematic one) of making and creating the very thing that is trying to be measured.

Counterfactual definiteness of art

The following is an explanation of Bell's theorem by showing, from the perspective of an artist, how he/she engages with capturing a reality: In an artist's deliberations when making an artwork, it is explicitly assumed that every possible mark (measurement), even if not performed, can be included in the visualization. The visualization involves critical reflection of the marks, averaging sets of outcomes that cannot all be simultaneously factual. If some of the marks are assumed to be factual outcomes of an experiential observation, others have to be assumed counterfactual. (Which ones are designated as factual is determined by the artist: the outcomes of the measurements he/she actually performs become factual by virtue of his/her choice to do so, the outcomes of the measurements he/she does not perform are counterfactual.) Bell's theorem proves that every type of quantum theory must necessarily violate locality or reject the possibility of reliable measurements of the counterfactual and definite kind. Local realism or locality presumes that an event happening at A cannot simultaneously effect B. The theory of relativity limits information to flow at the speed of light – it does not need to effect what happens to two entangled particles. If information is not being passed between two entangled particles then the laws of the speed of light are not affected.

On the World Science Festival YouTube channel, a video entitled *Measure for Measure: Quantum Physics and Reality* presents a discussion with four physicists defining their chosen methodology. Professor Rüdiger Schack presents a view of the quantum phenomenon of measurement that had been developed in collaboration with Carlton Caves and Christopher Fuchs. In Schack's session, he explains the new theory of QBism by using the analogy of a coin toss. By tossing a coin to make a decision, what is the probability of it being heads or tails? If someone was to choose a seventy per cent probability of being heads, and someone else was to predict a ten per cent chance of the coin being tails, then the likelihood of it being heads is greater than the proposed likelihood given to the coin coming up tails. Schack argues that the probability at stake here is a quantum linkage between belief and observation in matters of probability, one that deserves far more attention (Greene 2014).

Schack proposes a quantum link between states of probability and actuality, linking creativity and intuition with observation and determinacy. His claim is that the extent to which you believe what you will experience before the coin is tossed has some bearing on the outcome. This new concept in the quandary of measurement of the world at a quantum level has its roots in 1927 and in the conjecture of Heisenberg, thus facilitating an apt connection between QBism and Cubism. As a theory, QBism takes a theoretical possibility of dealing with the world that sees it as being different from what is constructed in the mind. What the coin analogy does is allow us to see our view of the world, but as individual and independent of the world of things. The coin is instrumentalized, allowing us to see its probability as an evolutionary tool.

The few artists mentioned in this chapter are synchronistically questioning systems of measurement, exploring the probability of what they are seeing and experiencing as to what is being represented in existence. The shackles of representing a mimetic world were lifted by the advent of photography in the 1820s. The question of what constituted the comprehension of the world was forging links with emerging scientific understandings. For example, the publication of James Clerk Maxwell's *A Dynamical Theory of the Electromagnetic Field* in 1865 brought together light, electricity and magnetism, and revealed them as the same phenomenon. Revelations such as this would surely have impacted Impressionist artists in that moment of shared discovery, where the concept of light as waves affected the surface of the canvas for painters such as Claude Monet.[20] In the same context, the concept of light particles is considered integral to the intuitive pointillist paintings of Georges-Pierre Seurat.[21]

The artist experiences the outcome of phenomena and brings it into a material existence, thereby rendering it from a quantum state to a single classical state of being. Yet it is not just the artist that does this – all human focused endeavours to image/render the world do this. This process is part of the continual conscious construction of the world as an ongoing manifestation through the critical searching for that intuitive becoming. Nonetheless, artists are exceptionally well positioned to direct their critical faculties and aesthetic interests into the making of worlds. The potential for the artwork to act as a becoming is at stake here, as will be discussed below.

Swerve speed

To define an artwork as being causally related to the swerve or the spin renders it as something inherently related to quantum computation,[22] which derives its power from the speed and superposition of the spin of particles. This speed can be seen as being comparative to the speed of thought. If the speed of a perturbance happening in the universe was faster than the speed of thought, then it would be a becoming as opposed to being. Deleuze and Guattari have discussed this becoming and the role of infinite speed as a defining element of chaos – it is the void, the vacuum energy field, the background radiation of the cosmos. I reference Michael James Bennett from his article 'Deleuze and Epicurean Philosophy: Atomic Speed and Swerve Speed' to aid in the articulation of what is meant by atomic speed:

> By retaining the Epicurean terminology and saying that the conceptual approximation of virtual infinity moves 'as fast as thought', Deleuze and Guattari imply that virtual infinity itself moves faster than thought, at 'swerve-speed' rather than 'atomic speed', so to speak. In both calculus and Epicureanism.
>
> (Bennett 2013: 145)

Here we can see how 'virtual infinity' made possible by quantum computation moves faster than the speed of thought. The swerve speed is Lucretius's clinaman, when atoms falling in the void unexpectedly swerve and from this unexpected swerve we get the cosmos, the planets and nature.[23] The swerve speed pre-dates being as such, and therefore should be, according to Bennett, 'understood as the speed of becoming' (2013: 145). Bennett suggests that being is a local and relatively stable system – a case of local realism – and not an 'essence' of being (2013: 145). Being is defined by its identity as a thing, by its 'identical' identity, whereas becoming happens prior to being. Bergson reiterates this problem here, writing:

> Now, that which lends itself at the same time both to an indivisible apprehension and to an inexhaustible enumeration is, by the very definition of the word, an infinite. It follows from this that an absolute could only be given in an intuition, whilst everything else falls within the province of analysis.
>
> (Bergson 1912: 7)

As Bergson points out, the concept of the infinite is linked with the inevitability that intuition must occur outside any frame of analysis.[24] Intuition is the very moment, the actual perturbance before some cognitive register moves from a state of not being known to a state of known-ness. The speed of intuition happens where actions or judgements are made beyond thought that happen automatically, as though possessed. This speed that reflects the infinite speed is in the realm of quantum phenomena. If, as we have established, we are quantum by nature and thought is atomic, then intuition can also be seen as a quantum procedure. Therefore, if the quantum computer works on quantum phenomena it could be said to operate at the same speed as conscious intuition. Quantum computation can tap into a speed that is beyond comprehension – so much so that metaphysics is seeping back into the conversation.

The Higgs boson,[25] or God particle, moves out of the stasis in the void, creating the slightest perturbance to take place in the fabric of the cosmos. This infinitesimally small motion happening at infinite speed is sympathetic with Epicurean atomism and the current discoveries in physics at CERN. The infinite speed needs to be enabled and engaged for intuition to happen in the physical world. If, during Lucretius's downward, vertical, parallel fall of atoms there was no disturbance caused by the slightest perturbation, a stasis would occur in the ineluctable repetition of the void. A perturbation would be the infinite speed of the unpredictable swerve, the diverted orbital spin that causes the miniscule change of direction. What are the conditions under which concepts of the world enter into concrete experience? It is the minimal change that effects parallel, endless falling of bits: the world as the history of the 'angle' in terms of the falling and colliding of atoms. So, here there are links drawn between a cosmological account of quantum activity – the 'swerve' – and the discordance of thought that we call creativity.

In her paper 'Cézanne and Lucretius at the Red Rock', Kathryn Tuma argues that intuitive dealings with atomic matter are at the very heart of Cézanne's work and practice. The position of Cézanne as a visionary, focusing on the assimilation and articulation of a new language so as to understand and visualize the world into existence, was intuiting Lucretius' poetic description that if it were not for the swerve 'Nature would never have created aught' (Carus 1950: 54). The swerve can then be seen at the heart of Cézanne's work (Fig. 6), not as an overt intention but as a simmering, niggling question fostered in his early scholarly studies. Cézanne's paintings resonate and link with the discussion of diagrammatic imagery in Chapter 2. His paintings qua diagrams plot a world that is based on an intuitive appreciation of uncertainty and limitless possibility. Cézanne's images give visual form to emergence and foreclose the illusion of the certainty of object-oriented ontology. This intuitive 'swerve speed' manifest in Cézanne's paintings is the necessary ingredient for the immense act of the artwork in attempting to bring the world into existence.

In his writing, Joachim Gasquet recalls a personal and anecdotal conversation that he had once had with Cézanne:

> [In order to paint the landscape correctly, first I have to discover the geo-graphic strata.] Imagine that the history of the world dates from the day when two atoms collided, when two whirling eddies, two chemical dances combined. These great rainbows, these cosmic prisms, the dawn of man over the void – I see these things rising. In reading Lucretius, I saturate myself by them.[26]
>
> (Cézanne 2001: 114)

In this passage, Gasquet likely elaborates and poeticizes a version of Cézanne's words, yet they still give us an understanding of that moment and the significance of Lucretius' writings when they were first read at a formative stage of Cézanne's life (Tuma 2002: 57). Can 'swerve speed' and the 'speed of thought' be seen as elements that are all used in the

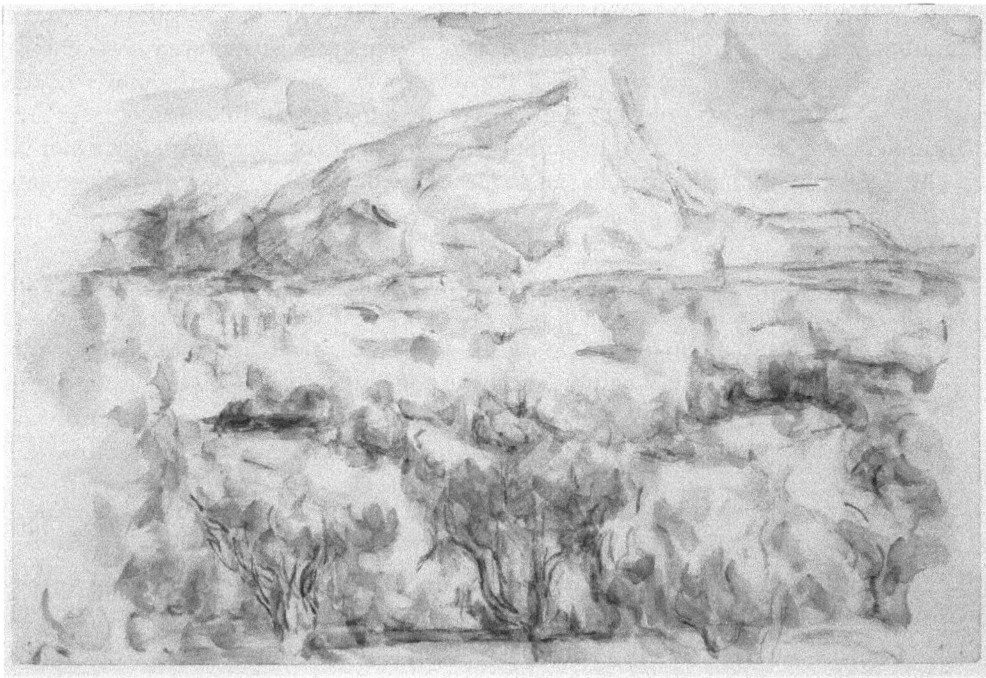

Figure 6: Paul Cézanne, *Montagne Sainte-Victoire*, 1905–06. Watercolour on paper. Courtesy of Tate Images London.

construction and observance of the physical world? In his letters to Herodotus, Epicurus explains the atomic world through an empirical comprehension, the lineage and provenance of which stretches all the way forward to Cézanne's modern musings. Epicurus stated in his time:

> For the assumption that beyond the range of direct observation even the minute times conceivable by reason will present continuity of motion is not true in the case before us. Our canon is that direct observation by sense and direct apprehension by the mind are alone invariably true.
>
> (Konstan 2014)

The model of the world that Cézanne was exploring links with the scientific discourse and discoveries that defined his time; new electromagnetic frontiers were opening the probabilities and possibilities evolving on the physics horizon, fundamentally altering how the world was modelled and understood.

At stake in this chapter is an analysis that spans the history of human thought in which, explicitly or incidentally, humans had set upon intuiting and controlling nature at a fundamental level by 'shaping the world atom by atom' and influencing the spin of the electron and nucleus of atoms (Amato 1999). The potential power that manifests from the twenty-first century state of this venture is exponentially growing and even now hints at the limits of our capacity to comprehend. No wonder that harnessing the power of the quantum is of interest, not solely to scientists and engineers, but to artists, writers and philosophers who see the ethical and political ramifications of quantum science and its capacity. Ascott questions the current role that classical definitions of 'nature' plays in this unfolding quantum and alien world. He writes:

> The shelf life of all metaphors, great and small is limited, and Nature seems to be past its sell by date. It has served us well in many respects – the ultimate appeal, for example, in questions of justice, human behaviour, as in those of representation and figuration in the arts has always been – is it natural? Thus the popular icon of the visual culture of the epoch has been Van Gogh (a value to be measured in hard yen) and, for higher brows, Cézanne, both of whom we are reminded spent most of their time in the fields, immersed in nature.
>
> (Ascott 2013: 437)

This scenario conveniently ignored Cézanne's comment that 'all things, particularly in art, are theory developed and applied in contact with nature' (Cézanne 1976: 228). An insight curiously parallel to that found in the new physics as exemplified in Werner Heisenberg's dictum that it is the experimental apparatus (which includes the observer's consciousness) that determines a particle's 'natural' behaviour. And it is from Cézanne that we first understand the consequences for painting of the artist's mobile viewpoint, his focus restlessly scanning a world in flux.

The world of flux is where we were positioned – now, to use the atomic flux as a quantifiable force is something that begs a moment of critical reflection on what Ascott and others have stated. The world as flux or constant vibration has been a way of thinking about the world, influencing artistic comprehension and the ongoing search for a reality. The flux-reality is now proven at a quantum level, a phenomenon beyond comprehension that nevertheless influences the way we see and think about the world. Now it is endemic of how we collapse the probabilities of all realities happening at the same time down to the one in which we live and breathe.

The quantum computer, governed by the spin at its atomic core that is analogous to thought, could therefore have the same potential of 'infinite speed' or 'swerve speed', not just the speed of thought. The speed of the thinking that leads to a new becoming has to happen faster than the speed of thought, which would then mean infinite speed or swerve speed. The quantum computer would then deliver the probability for intuition to take place as part of the processing capability.

The small perturbations are where we see the potential of the spin to form thought and therefore things. The epistemological evolution of the object has now led to the development of the quantum computer – a human mimetic device that could substantiate our own justification of intuition. In other words, the quantum computer is a device that has been developed based on our own instrumentalization of thinking and thought itself. The Epicurean construct of absence, of nothing being atomic at a fundamental level, means that at the microscopic level consciousness is atomic. The atomic performative agents in the quantum computer cannot move between the classical and quantum world in their unique states. In each state, both are real, both are physical and interdependent, but when confronted by a human mechanistic determinism we lose the opportunity for chance to confer life.

Quantum biology

With the idea of the quantum computer as analogous to thought, we can explore and unpack some expanding concepts taking place in relation to a field of study called quantum biology and consider if it might apply to art. Quantum biology explores whether nature uses phenomena such as tunnelling, entanglement and superposition of particles to enact the processing of energy and information with much greater efficiency. There is a direct relationship that can be made between the non-human atomic-quantum computer material processing capabilities and the materiality of the human brain. The human/non-human atomic connections need to be fleshed out by exploring relationships between quantum biological research and the mechanistic quantum computer research. The difference between the classical and quantum world is fundamental to the development of interest in the study of quantum biology. If we all come from a quantum past, then it would make sense that classical nature would use quantum phenomena to develop efficient ways of assisting in a biological evolutionary process.

At the end of his 1982 paper, Feynman joined the concepts of nature and quantum mechanics, stating that:

> I'm not happy with all the analyses that go with just the classical theory, because nature isn't classical, dammit, and if you want to make a simulation of nature, you'd better make it quantum mechanical, and by golly it's a wonderful problem, because it doesn't look so easy. Thank you.
>
> (Feynman 1982: 486)

To move forward from Feynman's works, the concept of our analogous relationship to thought is expressed in Matthew Fisher's paper 'Are We Quantum Computers, or Merely Clever Robots?' The paper builds on work being developed in the area of quantum biology and suggests that the brain itself functions as a quantum computer:

> Hopefully, if quantum processing is actually operational in the brain, it should be possible to co-opt the biological 'machinery' to duplicate the processing in a test tube – by analogy

> with genetic engineering. [...] If any *cross-correlations are detected* in the light emitted from the two test tubes it would be truly revolutionary, indicating the presence of nonlocal quantum entanglement between chemical reactions, perhaps paving the way for a liquid state quantum computer and for a quantum theory of the brain.
>
> (Fisher 2017: 9; emphasis in original)

What Fisher describes is the potential for the human brain and the quantum computer using the same phosphors atoms are one and the same. Some of the key terms such as quantum, tunnelling, entanglement, coherence and decoherence, used in describing quantum influences on nature, are based on this simple premise. If quantum is an effect of the universe, then why would it not be used by nature as part of the evolutionary processes in world creation? These specific concepts need to be elucidated to try to comprehend the plausibility of quantum actions actually having a significant impact on biological processes that at present can only really be intuited. How could these quantum elements now being realized and researched in biological systems determine the way that we, as humans, evolved?

The experimentation of quantum possible effects in photosynthesis via the movement of stable energy can be utilized as an example. In a paper published in *Nature Physics*, Neill Lambert and his collaborators suggest they present:

> [A] very brief overview of some of these cases in which quantum effects may assist or enhance a biological function. Our goal is to give a clear basic introduction to each system and to outline in what way quantum coherence or other quantum effects might be harnessed by this system.
>
> (Lambert et al. 2013: 10)

Photosynthesis possibly uses the effects of quantum tunnelling so as not to deplete energy whilst transporting it to the plant's reaction centre.[27] In photosynthesis, the speed of transmission without the loss of energy is needed in order for the plant to survive. This draws multiple parallels with how biological systems can work effectively when they 'let go' of a classical world view and adopt a quantum view.

Coherence is needed for tunnelling to be really effective. Coherence is equivalent to creating a quiet space where one can experience a singular event. For example, you are on your own in an art gallery looking at a particular artwork where the waves of light that are affecting you are constant and symmetrical. The spaces that you and the work are in form a spatial harmony. While the gallery is designed for this reflective behaviour, more people enter the space talking, taking photographs, listening to audio descriptions of artworks, carrying multiple hand-held mobile phones and generally disturbing the spatial ambience. The symmetry of the waves now leaving the artwork is disturbed and brutalized; coherence instantly changes to decoherence. The turbulence created from the waves seriously negates what appears to be a semblance of reality in that moment.

What quantum biology/physics gives us is indeterministic variation, the slightest perturbation that would change the repetition of information exchange. Entangled states in biological systems are extremely sensitive to noise, but something discernible in that noise can be the signal that you are trying to read. For example, when you hear your personalized ringtone in a busy noisy room, that particular frequency identifies that the call is for you. A personalised ringtone relies on that particular frequency being heard, to enable the transmission of information.

The complexity of life on a classical level challenges our ability to see the influence that could be taking place in the quantum zone. The complication of 'moist media' that constitutes life is when one can only take a holistic view of all the potentially different biological cells, neurons and particles communicating with each other simultaneously.[28] This holistic noise is very disturbing for the physicist's attempts to measure the probabilities of the effects of quantum phenomena. Quantum phenomena could be the very magic that explains something about life. It could be able to explain the inexplicable, but demands from individuals a suspension of belief in a purely classical world. Explaining the inexplicable would mean taking current concepts of reality by the scruff of the neck and allowing a more intuitive reasoning to adjust our perception of what constitutes being in the world.

The element of human intuition could be one such quantum-entangled phenomenon that enables many universal processes to be undertaken at the same time. If quantum phenomena are part of the evolutionary process at the core of the human condition, then we need to consider whether this phenomenon could have any real involvement with the human construction of art. What is the potential for the integration of the quantum mechanics in how we interpret the world around us? The world that artists want to capture and express is invisible and uncertain, hypothetically enabled by the role of intuition.

Intuition needs to happen using infinite speed and, following photosynthesis as one of the only means of processing information quickly in a biological system, quantum tunnelling. Then we could envisage the following: an electron's superposition could transfer information instantly between and through barriers that are not normally crossed by the classical processes of thought. This would enable particles that constitute thoughts to have the ability to exchange information in multiple places at the same time, therefore at swerve speed. Quantum tunnelling creates a rationale for intuition to utilize the noisy biological quantum world in femtoseconds.[29]

Does the possibility of quantum intuition help us understand what it might mean to be aware of the non-human/human concept of matter, such as the quantum effect on bird migration or a plant's photosynthesis? The main tenet of the concept of quantum intuition is the suggestion that uncertainty lies at the heart of all human endeavours; we may discover a certain beauty in recognizing quantum phenomena to be an integral part of the way we think about thinking. When art attempts to represent the unthinkable and incomprehensible, then at the very least we have to consider these forces as part an enlightenment process.

Notes

1 Michael Vaughan suggests that Bergson's 'élan vital' does not predict an immaterial force but, 'on the contrary, designates the vitality of matter itself, its organization, its growth, its indeterminacy, unpredictability and creativity, and this is inaccessible to mechanism in principle, not merely in fact' (Vaughan 2007: 16).

2 Hans Driesch was a biologist and philosopher whose experiments with embryology informed his theory of entelechy (a life-giving force). His work with organisms revealed to him that this life-giving force was an 'invisible presence' that could not be performed by a machine (Bennett 2010: 61).

3 In this context, Karen Barad explains that 'matter is not a fixed essence; rather, matter is substance in its intra-active becoming – not a thing but a doing, a congealing of agency' (Barad 2003: 80).

4 Étienne-Jules Marey was the epitome of a transdisciplinary researcher; he was a physiologist, doctor, biomechanics engineer and inventor whose work was significant in developing a number of physical instruments used in the areas of medicine, aviation and cinematic and photographic exploration.

5 Ether was a hypothetical medium that at the beginning of nineteenth century was put to rest. Isaac Newton who first used the term as early as 1781 thought of it as a substance to fill all space and later James Clerk Maxwell saw it as the medium to transport light through the propagation of electromagnetic waves.

6 This reference shows that there was disagreement about the value of Marey's work, and Boccioni sided with Bergson over the camera's instrumentalizing of time and space (Braun 1992: 279). 'Marey's friend and colleague at the Collège de France, Henri Bergson, was equally dismayed at the way chronophotography reduced time to calculable space and saw Marey's work as contributing to the subordination of human experience to an objective and external measure' (Corbett 2008: 13).

7 Boccioni drew heavily on Bergson's concepts of matter. Jack Burnham states that '[v]italism in sculpture can be viewed as a prototype of the trend toward stylistic transcendence. It attempted, at the beginning of the century, to conjoin idealism to a new awareness of biological capability. In one sense, vitalism stood not even halfway between the two sensibilities; it took a step – a large one – out of the old world of representational idealism. A source of intense belief was needed, something that would replace fast dying faith in realism. Because of his timely and successful attacks on materialism, Henri Bergson became the high priest of a new cult – that of instilling life, by way of the vital essence transferred by the sculptors fervor, into inert matter' (Burnham 1968: 70).

8 The superposition is described as being one where an electron can be in multiple states of being in all possible positions at the same time. This is not the same as being, in either-or positions, which is how a classical interpretation of the world works.

9 Henri Poincaré was a French mathematician, theoretical physicist, engineer and philosopher of science who was strongly linked with influencing the artists of his time, most notably Pablo Picasso's link to four-dimensional space. In Poincaré's chapter on Classical Mechanics from his book *La Science et l'hypothèse*, there is a critique of an experiment by Jules Andrade

from his publication *Leçons de mécanique physique*. In this experiment, a piece of thread is used to measure and explain velocity, which is indexical to Duchamp's *Three Standard Stoppages* (Poincaré 1905: 107–10).

10 We can imagine that of particular interest to Picasso was Poincaré's suggestion of how to view the fourth dimension. Images from the fourth dimension can be projected onto a three- or two-dimensional surface from different points of view or perspectives (Miller 2004: 488).

11 Henderson summarizes the complexity of writing about the holistic environment that affected the artist's imagination: 'Borrowing Hayles' idea of the "cultural matrix" and anchoring it historically by analyzing available popular sources on science, we will never know all that the matrix contained, but we can be fairly certain of what it could not contain. In pre-World War I Paris, that was accessible information about Einstein and Relativity Theory. Popular scientific literature, along with the writings of artists and critics, is ripe for the discourse analysis Mermoz following Foucault, proposed' (Henderson 2004: 438).

12 Einstein wrote to Max Born and references 'God who plays dice' on 7 September 1944, stating 'We have become Antipodean in our scientific expectations. You believe in the God who plays dice, and I in complete law and order in a world which objectively exists, and which I, in a wildly speculative way, am trying to capture. I firmly believe, but I hope that someone will discover a more realistic way, or rather a more tangible basis than it has been my lot to find. Even the great initial success of the quantum theory does not make me believe in the fundamental dice-game, although I am well aware that our younger colleagues interpret this as a consequence of senility. No doubt the day will come when we will see whose instinctive attitude was the correct one' (Einstein et al. 1971: 149).

13 The Copenhagen Interpretation suggests that the wave function is a complete description of a wave/particle. The wave function collapses when a measurement of the wave/particle is made. If there is uncertainty in the relationship between two objects, there is no way of measuring them simultaneously. If we measure a wave/particle's position, its momentum becomes uncertain.

14 In the conclusion of Bell's 1964 paper 'On the Einstein Podolsky Rosen Paradox', he states: 'In a theory in which the parameters are added to quantum mechanics to determine the results of individual measurements, without changing the statistical predictions, there must be a mechanism whereby the setting of one measuring device can influence the reading of another instrument, however remote. Moreover, the signal involved must propagate instantaneously, so that such a theory could not be Lorentz invariant' (Bell 1964).

15 In the Born–Einstein letters based on the correspondence between Albert Einstein and Max and Hedwig Born, we find the quote: 'I cannot seriously believe in it because the theory cannot be reconciled with the idea that physics should represent a reality in time and space, free from spooky actions at a distance' (Einstein et al. 1971: 158).

16 Quantum Bayesianism or QBism relates to an agent's actions and context as a fundamental theory in quantum mechanics. QBism is noted by its use of a subjective Bayesian proposition of probabilities to comprehend the quantum mechanical through a simple set of propositions.

17 Freud's work *The Ego and the Id*, first published in1923 in German, was translated into English and published in 1927 (Freud 1927).

18 Atomism, developed by Democritus (460 BC), is a theory that atoms are the smallest, discrete and indivisible elements, which are the ultimate constituents of all matter.

19 Charles A. Hobbs defines Peirce's relationship with science as informed by 'tychism, an ancient view that is most striking in Epicurean thought. This view comes from the Greek (tyche) for "chance" or "luck," and refers to the hypothesis that the universe is fundamentally indeterminate, or that, contrary to the doctrine of necessity, there is no law of nature that is completely exact or unchangingly precise. Such a doctrine, according to which there are immutable laws, cannot be viably maintained even just as a presupposition or postulate demanded according to the needs of science' (Hobbs 2008).

20 Claude Monet's Rouen Cathedral series of paintings created in early 1890s could be seen to explore the possibility of waves effecting matter: 'The ether, which supported the wave theory of light, invites a rather differently nuanced reading of painters' views about their visual experience of the atmosphere, their desire to represent the actual mediums that enveloped objects, filled space, and transported light. The ether evoked the idea of a physically palpitating, coloured "materiality" out there to paint' (Callen 2000: 188).

21 Seurat's painting *A Sunday Afternoon on the Island of La Grande Jatte* (1884) constructs a window through which the distance between us and the world of atoms is calibrated to give the world presence. We can understand Seurat's work as picturing light behaving like particles, operating in a void scaled up for our perception.

22 Quantum computing makes direct use of quantum-mechanical phenomena, by utilizing such attributes as superposition and entanglement of electrons, nuclei and photons, to perform operations processing data.

23 Bennett suggests three interpretations of Deleuze's two concepts of the velocity of atoms:

> (the continuous speed of thought and the faster speed of the swerve); there are three (the continuous speed of opinion, the discontinuous speed of thought, and the even faster speed of the swerve). What Deleuze explicitly says – that the swerve, as a determination of the relation between atoms, is accomplished faster than a minimum of continuous time – is true but trivial, since for Epicurus the continuous time of opinion is slower than even the time of thought. But what Deleuze wants to say is far from trivial: that the determination of a relation among atoms is accomplished even faster than the smallest time determinable by reason – faster than the true minimum of time, not just faster than opinion but 'faster than thought'.
>
> (Bennett 2013: 138)

24 For Bergson, the concept of absolute is understood to be an infinite 'held as relatively to other things, [...] the gold coin for which we never seem able to finish giving small change'. The absolute could then only find its place in the realm of intuition where one sees oneself 'within an object in order to coincide with what is unique in it and consequently

inexpressible'. The inexpressible object is the part that needs to be comprehended but can't be; it is the state of a non-reality that moves between the quantum and the classical. The space of intuition is where one experiences something that is asked to be believed without proof of its reality. We use the 'province of analysis' to suggest that there is some way to agree to the nonreality to substantiate what can't be proven (Bergson 1949: 7).

25 The Higgs boson (named after Peter Higgs) was first suspected in the 1960s and was discovered on 4 July 2012 through experiments in CERN's Large Hadron Collider.

26 The first line of this quote is referenced directly from the official translation of Joachim Gasquet's conversations with Cézanne. The remaining lines of the quote are taken from Kathryn Tuma's translation of the original French (Tuma 2002: 56).

27 Quantum tunnelling is where you try to get from point A to point B through some material that would not normally seem possible. It is always a faster movement of energy than the classical model of information hopping from one atom to another. For example, we have a wave as a particle – as the wave moves, it leaks through the material that is blocking its path. The particle wave starts out with its probable position being at the peak. As the wave moves through the material, the particle is in both places at the same time, rising to a peak on the other side. As the process takes place instantly, the particle wave has now moved through to its peak on the other side of the material with the same energy intact.

28 'Moist media' is a term coined by Roy Ascott to deal with the inclusivity of the biological arts within contemporary art thinking: 'Between the dry world of virtuality and the wet world of biology lies a moist domain, a new interspace of potentiality and promise. I want to suggest that (comprising bits, atoms, neurons, and genes) will constitute the substrate of the art of our new century' (Ascott 2000: 1).

29 In the *Oxford Dictionary* a femtosecond is defined as 'one quadrillionth of a second' (Dictionary).

Chapter 2

The Diagram

Convergence

For every unfolding of a phenomenal event, there exists a multitude of ways of describing it. Descriptions, models and mappings are formal articulations of our fundamentally uncertain relationship to the world: uncertainty that is constituted as (and of) thought. The material forms of thought-expression (oration and the spoken word, bodily and embodied gestures, social actions, symbolic forms and textual production) have all been reified by centuries, if not millennia, of human history – and now serve us, but with an exaggerated sense of security (of knowledge), as everyday vehicles for the articulation of immaterial thoughts.

The scientific diagram, one of the many expressive forms of human consciousness, is a systemized response to, and of, the visualization of complexity. Scientific diagrams are a demonstrable example of the latent human capacity for the systematic transference of meaning between conscious minds. The historical development and proliferation of the diagram (and thus its significance) is a broad and multifaceted one – and therefore exactly what constitutes a diagram in the scope of material expression is fuzzy. The focus at this juncture and in this chapter is to explore the visual representation of invisible thoughts (consciousness) through the medium of the diagram.

A diagrammatic gesture is one that expresses and marks the shape of thought. In so doing, the body of the mark-maker is exposed to an evocation and concretization of conscious (creative) expression – one that is thus generated to translate ideas into a formal image of the system and structure of the world. In this context, diagrams are a series of marks, points and lines that serve to articulate a process of thinking and feeling 'through' the world and always in direct correlation with the human-centric desire to express the processes of our consciousness. The expression of the mind onto a material substrate enables a conceptual projection to become manifest, as it inscribes the very matter that the mind seeks to make sense of.

A telematic process of 'revealing' is at the heart of all world-modelling: the mapping of our phenomenal and material surrounds. By mapping in this way, (otherwise) invisible threads of what appear to us as 'meaning' are allowed to manifest, to become visibly inscribed upon a material substrate that becomes a membrane between the world and us. It is this yearning of humans to mark and express that renders all forms of cultural-conscious expression as linkages between a thinker and the material world. Commenting on this phenomenon of the diagram, Hélène Frichot writes:

> The work of the diagram in the philosophy of Gilles Deleuze describes a sometimes violent exertion of thought in the face of the unthought, which is at the same time a battle between finitude and the infinite, or a struggle between (human) processes of 'subjectivation' and the forces of the outside. By stressing the term 'subjectivation' Deleuze wants to emphasise the mobile and ever-transforming emergence of the human subject in relation to diagrams of power.
>
> (Frichot 2011: 1)

Frichot notes the social dimension and effect of the diagram that collapses possibilities of meaning, through gestural expression. Here, the complexity is one of articulating the unthought that is to be made real, which is as much a part of the process as the thought that is being made transparent in the making of the diagram.

Diagrammatic mechanics reveal an inner-working exemplified in the (diagrammatic) doodles of the artist Paul Klee. Klee's artistic marks appear to be synchronistic with cosmic forces in terms of the parallels between how universes and worlds are brought into existence from a single point of energy, and how creative works expand outwards from a single graphic vector. The artwork here becomes analogous to the diagrammatic process of working from subconscious activity of the mind (intuition) and unfolds with remarkable and unpredictable complexity, loaded with probabilities for 'becoming'.

In developing this link between creative expression, cosmic 'becoming' and diagrammatic imaging, we turn here to a well-known diagram created by the theoretical physicist Richard Feynman. Feynman's diagrams became a signature mark of his thinking process and are thus, given his importance in the articulation of quantum mechanics, well-known and commonly referred to. Feynman's diagrams are unique visualizations of the atomic-quantum world. Because Feynman sought to image those properties that constitute the quantum world, properties so alien and strange to our common sense, the critical reception of Feynman's diagrams is an important part of the story.

The diagram was a way for Feynman to actively understand, explain, innovate and develop schematic research for thinking through complex problems being encountered in contemporary physics. Feynman's diagrams came from a spark of imagination and alternative thinking processes to recognize, explain and articulate, even as a simple line, the phenomena taking place within the quantum world. The line work in Fig. 7 is significant because it shows the indecision taking place in the mind as it is transferred to the line work in the four small diagrams at the bottom of the page of notes. Here, the marks sometimes cancel each other out in an attempt to communicate across multiple fields of research.

To explain the significance of the line to the phenomenon of thinking, we can consider the practice of hand-written signatures as an example and expression of a personal diagram of the self. Written signatures can be interpreted through an analysis of their significant gestures: arches, dips, angles, peaks, troughs, sweeps, soft, hard or sharp movements and swirls of gestural lines. These provide an insight into the complexity of a complete individual's profile. The written signature is therefore a line revealing characteristics that, when unfurled,

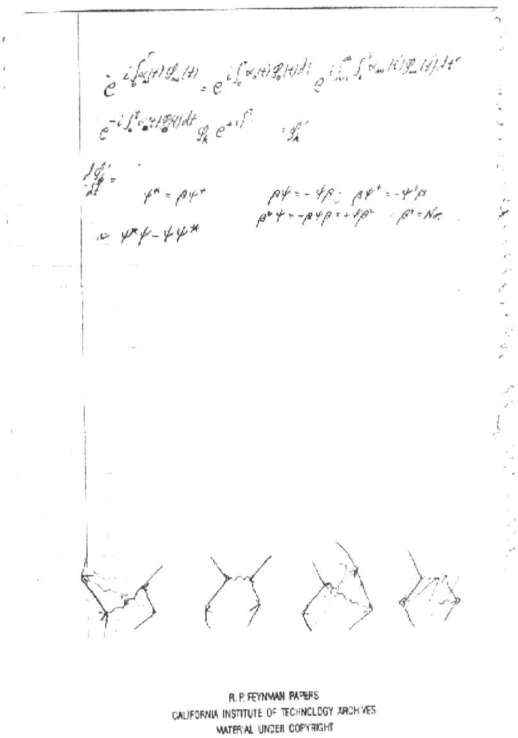

Figure 7: Richard Feynman, Sketch and notation. Courtesy of the Caltech Archive.

return to the universality (non-character) of the straight line. The meaningfulness coded in the expressive capabilities of a hand-written signature is diagrammatic in its nature, but it is not rigid – rather, it is open to many interpretations.

Diagrams often attempt to describe and present something that is different from anything encountered or experienced before within our field of reference. The reading of the line as having more to offer than a simple recognition of the person's legal identity through the signature allows for the viewer to express an abstract concept in the interpretation of the mark. The diagram can thus become a point of rupture within scientific discourse: a creative rupture at the site of the means of expression in scientific inquiry.

In Fig. 8, we see Feynman's drawing of the back of someone's head, most probably produced during a conference presentation. I speculate from this drawing that the hair is seen by Feynman as being diagrammatic, a starting point for other ideas to develop on

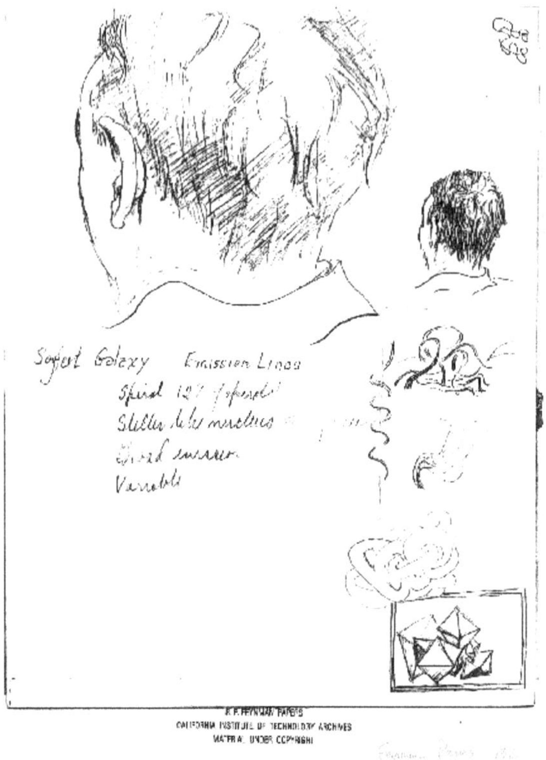

Figure 8: Richard Feynman, Sketch and notation. Courtesy of the Caltech Archive.

the page. He makes the connection by focusing on a detailed study of the formation of an individual group of hairs. The accompanying text references the Seyfert galaxy that is typified by an intense bright core that presents strong infrared emission. The hair is further deconstructed to reveal a detail, almost indicative of a single helix. The five small drawings to the right-hand side of the page show a progression of the hair towards geometric abstraction, reminiscent of Picasso's famous *Bull* lithograph series (1945).[1] In the bottom right two sketches of hair (Fig. 8) are firstly transformed into interlocking chain links, while the second transforms the chain into what appears to be a rhombus.

In Feynman's diagrams, the necessary invisibility of quantum mechanics is made quantifiable as a diagrammatic form of expression. Our understanding is geared so that we want to see quantum physics expressed in ways we are familiar with, as resembling something we already know. Feynman understood this, yet when we propose a simple solution of mimicry and anthropomorphism – familiar visual tropes borne of the human condition and human-centeredness in thought – we produce only an emptiness of pictorial resemblance, that is a reiteration of the symbolism that we already know.[2] Feynman used visual thinking (a necessity) to work through his ideas in developing a new visual-diagrammatic language through his (now) iconic diagrams that reference not certainty, but latent probability as fundamental to quantum mechanics.

Diagram and the new language

In his book *Drawing Theories Apart: The Dispersion of Feynman Diagrams in Postwar Physics*, David Kaiser demonstrates a similar problem to that faced in artistic expression when confronted by the need to cultivate a new visual language through which to express and explain how the world truly appears. He writes:

> Today physicists all over the world scribble down the diagrams when studying everything from the mundane to the bizarre. Whether calculating the properties of materials that will form electronic computer chips, the levitation of magnets above superconductors, the behavior of particles near black holes, or the origin of matter itself just fractions of a second after the big bang, physicists begin by drawing Feynman diagrams.
>
> (Kaiser 2005a: 4)

In the 1947 Shelter Island Conference, Feynman first introduced his quantum diagrams to a perplexed audience, and again the next year at the Pocono Manor Inn (1948). No one at the Shelter Island Conference seemed to be able to follow Feynman's presentation, which came after a long day of presentations. The physicists in the audience could not comprehend what Feynman was doing or how his new un-scientific doodles fitted in with the more established historical principles of quantum physics (established in the 1920s by Max Planck). As stated, the quantum-mathematical problems that Feynman was trying to express were highly complex and challenging, due to the sheer amount of mathematical calculations required to factor in when dealing with, and imaging, quantum probabilities.

The quantum perturbative calculation was the method used to try to eliminate some of the less important calculations in quantum mechanics in order to deal with the calculations that made the largest contribution to the problem at hand. Kaiser makes sense of this problem in his book; he writes:

> Looking back, we might expect the reception from his colleagues at the Pocono Manor Inn to have been appreciative, at the very least. Yet things did not go well at the meeting. For one thing, the odds were stacked against Feynman: His presentation followed a marathon day-long lecture by Harvard's Wunderkind, Julian Schwinger. Schwinger had arrived at a different method (independent of any diagrams) to remove the infinities from

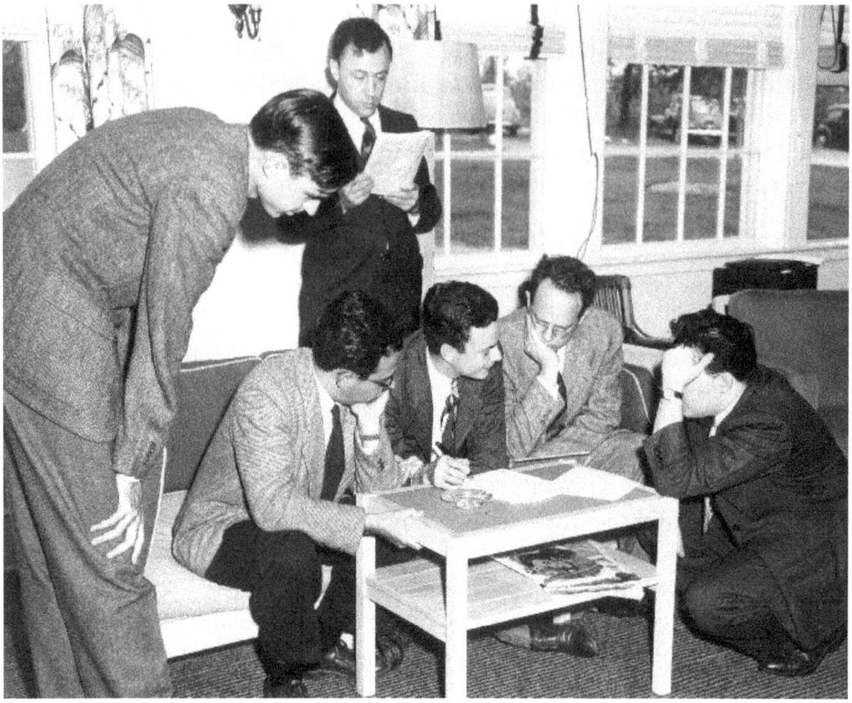

Figure 9: Group at the Shelter Island Conference. Left to right: Willis Lamb, Abraham Pais, John Wheeler, Richard Feynman, Herman Feshbach and Julian Schwinger. Courtesy of the Emilio Segrè Visual Archives, American Institute of Physics.

> QED calculations, and the audience sat glued to their seats – pausing only briefly for lunch – as Schwinger unveiled his derivation. Coming late in the day, Feynman's blackboard presentation was rushed and unfocused. No one seemed able to follow what he was doing.
>
> (Kaiser 2005b: 160)

Citing Heisenberg's uncertainty principle, for example, Niels Bohr raised repeated objections to the very notion that spacetime diagrams could be of help for studying what 'Bohr insisted were inherently unvisualizable quantum phenomena' (Kaiser 2005a: 47). The comment by Bohr would suggest that intuition as opposed to scientific reasoning is necessarily the first point of contact with understanding the potential for exploring the visualization of the unvisualizable. We will explore this problem at a later point through the notion of the infinite speed of thought.

In examining Fig. 9 we see the men with their hands on their heads or mouths, presumably deep in thought, yet John Wheeler showing no interest at all. Willis Lamb stands at arm's length in these cramped conditions, which were obviously not suitable to express such complex concepts of abstraction. The photograph raises connections to the similar disbelief in the art world when new visual forms of expression do not correspond to the 'norms' of representation.

The line

Feynman shows that graphic lines are capable of expressing a seemingly inherent and unvisualizable quantum-ness. This re-envisioning of the capabilities of the line shares something with the term entelechy, 'living vitality', which can be seen as a realization of actuality (as opposed to potentiality). This vitalism of entelechy is exposed in the work (and writing) of many artists – but here we will limit the discussion to two Bauhaus artists, writers and educators: Paul Klee and Wassily Kandinsky. In Klee's 1925 *Pedagogical Sketchbook* (published the year before Kandinsky wrote *Point and Line to Plane*), the author explores expressive qualities of lines in terms of how line-work may coerce consciousness and understanding.[3] Klee and Kandinsky[4] were writing on this topic to illustrate principles associated with line and form during their time at the Bauhaus in Germany. There, both artists sought to unpack what they mutually perceived to be a deep and causal relationship between physics and cosmology on the one hand, and the line as a fundamental form of expression on the other. It was this cosmological line that configured and conveyed a qualitative experience of being in the world.

The diagrammatic drawings in Klee's book are attempts to explore the spectrum of possibilities that exist in communicating between human agents via the proposed universal expressiveness of lines.[5] Through his theorizing, quantum worlds seem to be at the heart of Klee's work, leading from the recognition of the power of the diagrams that reference the forces at work in the relationship between mind and (graphic) line. The point that activates the line with the initial touch or the mathematical point on the perspectival horizon, where converging lines terminate, both function in the same way, according to Klee:

> From point to line. The point is not dimensionless but an infinitely small planar element, an agent carrying out zero motion, i.e. resting. Mobility is the condition of change. Certain things have primordial motion. The point is cosmic, a primordial element. Things on earth are obstructed in their movement; they require an impetus. The primordial movement, the agent, is a point that sets itself in motion (genesis of form). A line comes into being. The most highly-charged line is the most authentic line because it is the most active. In all these examples the principal and active line develops freely. It goes out for a walk, so to speak, aimlessly for the sake of the walk.
>
> (Klee 1961: 105)

In the year after Klee's book was published, Kandinsky wrote *Point and Line to Plane* (1926), wherein he explores the intrinsic forces and expressive qualities in the graphic line. Both

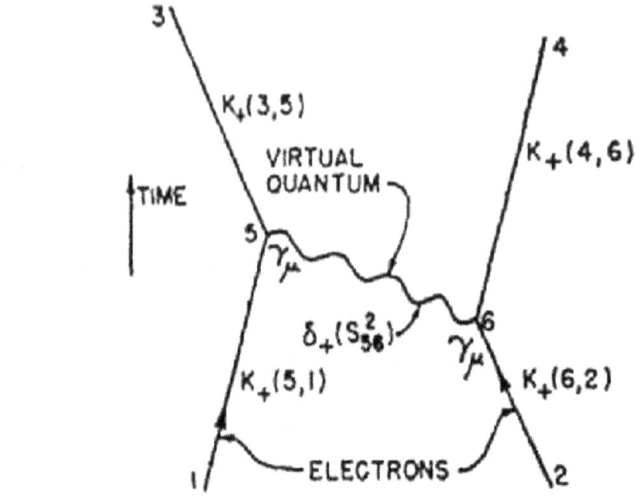

FIG. 1. The fundamental interaction Eq. (4). Exchange of one quantum between two electrons.

Figure 10: The first published Feynman diagram appearing on page 772 of Feynman's paper 'Space-time approach to quantum electrodynamics' published in *Physical Review*, 76: 6, pp. 769–89 on 15 September 1949.

Klee's and Kandinsky's analyses share the idea that the line has characteristics that are charged with a particular and flexible agency. It is the same agency that is subliminally constituted in the line work present in Feynman's diagrams. In other words, Feynman, Klee and Kandinsky are co-constituted in the struggle to rationalize some primal source of meaning of representing phenomena in the world: the graphic line.

To return to Feynman here, the diagrams that he presented in his lectures are different to the diagrams seen above (Fig. 10). In the performative drawings-cum-diagrams he employed in his lectures, Feynman demonstrated how those graphic chalk lines he produced were, in fact, vital agents for exploring and making sense of physics' scientific concepts.

Through line, Feynman was able to graphically articulate his thinking so as to present the fundamental quantum probabilities of photonic behaviour. Through this comparison, we see a vital (and vitalistic) contrast between static versus performed/gestural diagrams, which bridges the gap between the diagram as the repository of static knowledge versus the artwork as the active and lived/felt engagement with conscious existence.

The diagram pictured in Fig. 11 acted as a catalyst for the development of a body of work concerning developments and imaging in quantum phenomenon and thus necessarily (if at once unconsciously) informed the subsequent development of this book. Feynman's 1979

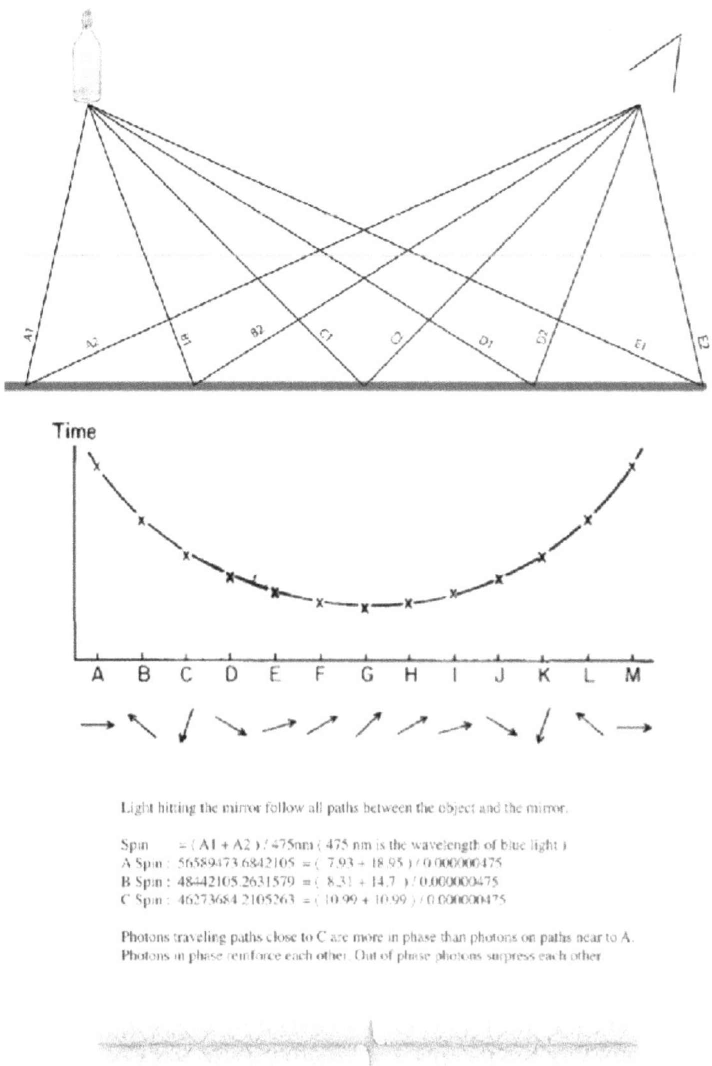

Figure 11: Reworking of Richard Feynman's diagram for the work *Multiverse*, 2010. Courtesy of artists Paul Thomas and Kevin Raxworthy.

performed/gestural diagrams became instrumental in the elaboration of a media artwork, entitled *Multiverse* (2010), created by a collaboration between Kevin Raxworthy and myself. The research undertaken during the creation of *Multiverse* examined and expanded Feynman's data to create an aesthetic translation that, surprisingly, when transposed into an artwork depicted the probability of a parallel universe. These unexpected results of the artwork will be discussed in more detail to consider how the visual phenomenon created drew a synchronistic relationship with the use of the diagram within the methodology of another diagrammatic artist/painter, Francis Bacon, whose work we now turn to (Deleuze 2005: 99).

The flexible diagram

The expressive power of Feynman's diagrams in visualizing the photonic world by their probabilities and in conjuring a graphic 'becoming' is one that can be brought into focus through a comparison with the gestural paintings of Francis Bacon. Bacon's paintings derive their remarkable expressiveness via the artist's use of diagrammatic demarcations to present a series of both visualized and actual (that is, gestural) 'becomings', which are what we perceive in his finished paintings. To stress this point, we are conceptualizing the creative work of Bacon as using a method of painting that is diagrammatic, an idea substantiated by Deleuze (2005: 99). We can conceive of both Feynman and Bacon as drawn towards the functionality of the diagram as a means to explore the probability of something occurring and to image, or give form to, that occurrence.

The diagram, we can now say, is a framework that has been utilized (if only intuitively) as a means for artists and physicists alike – a framework that undercuts both discourses. Through the diagram, Feynman contextualizes in visual form the phenomenon of Quantum Field Theory (QTF) where particle excitations happen within a field, by invigorating his graphic lines with a back-and-forth movement. To make sense of how this technique worked in conveying something of the quantum world, we might imagine the physical world as a sum of fields, and the sum of fields as an entangled mesh. Like a fabric produced from the spinning of yarn, we can thus see a number of connections taking place between seemingly distinct locations. These invisible operations are the same quantum effects that Feynman was attempting to visualize. If we run with the analogy and maintain the stretched canvas (fabric) as a tabula rasa of invisible threads, then we can conceptualize the canvas as a void of expectation. Like Feynman's chalk marks, Bacon's painterly gestures served to construct and develop his initial subconscious marks-cum-vectors. The graphic functions expressed by both of these men were bound up in the same physical phenomenon: graphic excitations made on the physical 'fabric' (like the weaving of a signature) in such a way as to imbue their marks with the very sense of their own becoming (an ontology of the line).

Therefore, Feynman's diagrams sought to explore invisible forces that would, when revealed, serve as a visualization that would (as shown) be probably unfathomable to most people in the 1940s. The diagrams demanded intuition, expression, recognition, perpetuation, courage and conviction in order to be read as they were intended – they demanded a shift in perception. To further explore the fundamental and binding relationship between Feynman and Bacon, we need to look closer at what physical processes underwrite their respective work. Namely, we must examine how visual complexity relates to the character of the quantum at the classical scale of intuition and objects. In other words, we must examine how the work of Bacon and Feynman essentially serve to 'scale-up', whether from quantum mechanics or from consciousness, so they can be made visible and thus intelligible in the realm of the everyday and its classical *modus operandi.*

The difficulty with scaling quantum phenomena up to the everyday realm of objects is that our intuition, or consciousness, is bound up to the scale of this quotidian-classical field. Even when we scale-up quantum mechanical systems or phenomena, they don't make intuitive sense to us. The quantum world is one where things exist in multiple states at the same time. Furthermore, quantum phenomena exclusively function at the quantum scale; any attempts to scale-up quantum states result in fixed states where they are torn from their 'natural' quantum habitat. Thus, the classical scale and the quantum scale are seemingly incompatible. Therefore, the job of image-makers and diagrammers of quantum mechanics is to convey (that is, to not lose) the irrational complexity of quantum states while making them intelligible at the everyday scale of perception. The artist and scientist are confronted with this phenomenon constantly on account of the nature of their engagement with the world. Creative thinkers are required to interrogate the conceptual framework of reality in order to resolve questions about the 'what' and the 'why' of a world.

Feynman contended that nature is quantum, that the seemingly irrational complexity of the quantum world manifests into and underwrites the world of objects and, by necessary extension, consciousness and perception (1982: 486). The very nature of being human is one that is, and has always been, quantum at its very core. It is this quantum complexity that can be seen in art that attempts to express what it is like to be present in the world, even though it is only recently that quantum qualities of expression could be understood (and named) as such. Bacon, Klee, Kandinsky: these artists sought to cultivate a perception of the world that was bound up in the invisible influences of, and being present within, quantum-nature. Using the word 'present' here I mean it in terms of an inquisitive state of conscious questioning, thereby a sense of creative awareness capitulated in Marshall McLuhan's adaptation of Ezra Pound's conception of creative thinkers/artists as 'antennae' that serves to tap into the invisible energies and transmissions of the cosmos.[6] Being 'present' then is a term for psycho-social intuition in which artists and creative enquirers channel the world much like the delicate needle of a seismograph channels tectonic murmurs. It is worth citing McLuhan in full:

> The power of the arts to anticipate future social and technological developments, by a generation and more, has long been recognized. In this century Ezra Pound called the artist 'the antennae of the race'. Art as radar acts as 'an early alarm system,' as it were, enabling us to discover social and psychic targets in lots of time to prepare to cope with them. This concept of the arts as prophetic, contrasts with the popular idea of them as mere self-expression. If an art is an 'early warning system,' to use the phrase from World War II, when radar was new, art has the utmost relevance not only to media study but to the development of media controls.
>
> (McLuhan 2003: 16)

Signals from the non-human world thus breach the human world through the 'antennae' of artistic and creative present-ness; a state of ontological inquiry that processes and incorporates non-human transmissions, and internalizes, creates and presents them to the human world (consciousness). The artist gives a sense of the present through their practice that may foretell, express or evoke future states (serving as revelations). The 'revelatory' function of creative expression in the graphic line (and elsewhere) in terms of the ability to visualize and make intelligible quantum phenomena is what is at stake in this book.

The state of present-ness and temporal significance can be seen in the work of Kazimir Malevich's painting *Black Square* (1915) that is a moment of recognition of the uncertainty of the void beyond representation.[7] Painted in a seemingly darkened room, it is an intuitive example of the void coming into existence within the arts at a time when physics is grappling with its own burgeoning comprehension of concepts of space and time. The diagrammatic *Black Square* is an ultimate summing of possible realities, a painting that comes a year after Duchamp's *3 Standard Stoppages* (1913–14). The materiality of the blankness develops in contrast to the representation of the measurement as the quantifiable and definitive. This void is where the fall of the Epicurean atoms takes place; it is where the uncertainty in the world begins.

There is an ongoing duty to challenge the robustness of physics and science as being the static arbiter of knowledge of the physical world. Art that questions world-models and how they emerge can be said to maintain its unique quantum phenomena of nature. The fundamental challenge is for us to see the role of the artist as 'antenna' in relationship to the transmission of a full spectrum of signals. The signals are inclusive of human and non-human elements within the universe. The signals-as-agents are shaped by the human sensory mode to be recorded and then acted upon to the creation of new cultural transmissions.

The speculations of Newton in explaining aspects of light, heat and ether show the very fragmentation of the world via the triggering of thoughts; through the acts of witnessing, receiving and processing transmissions of the world around him. The translation of these fragmented transmissions is studied as probable answers to the observer's conundrum, that is trying to find oneself in a dynamic universe. Here we can imagine the world as a radio whose tuning receiver is scanning for stations and which picks up fragments of semi-known languages, forming a strange staccato nonsensical sense. The artist as 'antenna' needs to tune the signal that is being picked up to make sure that they are not transmitting a memetic message or just plain noise back to the world.

In his art practice, Richard Goodwin uses equations and diagrams from physics in coating his three-dimensional objects and on the surface of his paintings and drawings.

This is evident in the work *Moth* (Fig. 12) where the surface of a beautifully varnished wooden boat is covered with the equations and diagrams from historically significant physicists. These elements are randomly chosen and therefore have a certain probability of being, and not being, meaningful. The elements are sourced from books such as *Analytical Experimental Physics*, *A Guide to Feynman Diagrams in the Many-Body Problem* and *Calculus and Analytical Geometry* (Ference 1956; Mattuck 1976; G. B. Thomas 2003).

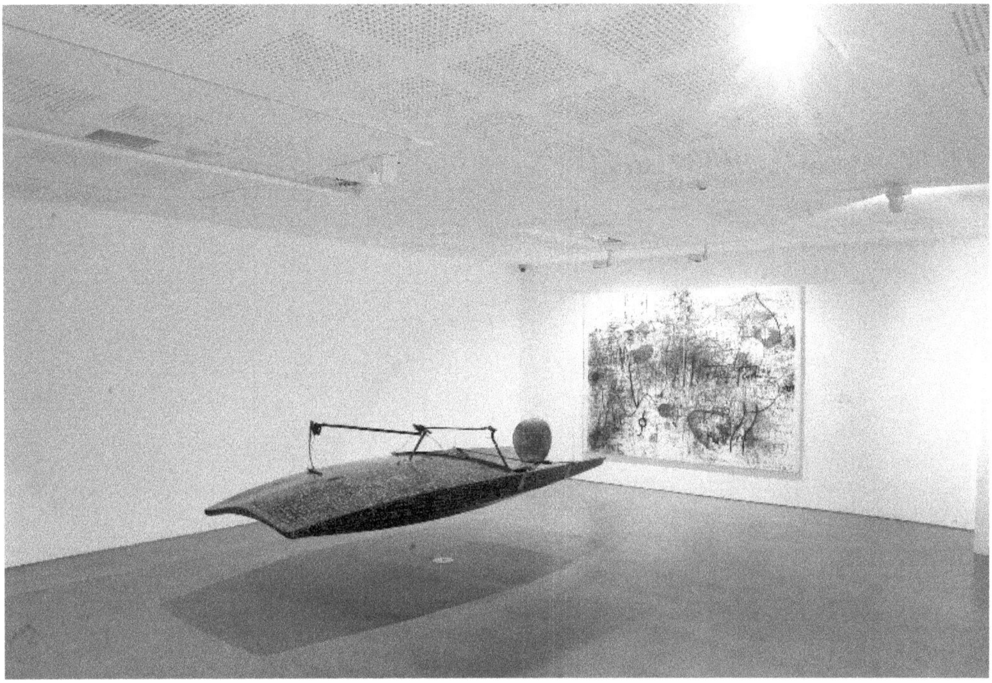

Figure 12: Richard Goodwin, *Moth* (installation). Courtesy of Richard Goodwin.

The equations and diagrams transform (Fig. 13) the boat's skin into a living example of the latent mathematical probability within all things. The boat that bobs up and down on the waves emulates the probable position of where a particle could be found. The artwork transcends the elements into an allegorical object that coalesces the classical and seemingly counter-intuitive quantum worlds in which we live. The study of object-ness as illusions echoes the famous 'digital rain' in the science fiction film *The Matrix* (1999) that transforms and blends the human's incredible ability to comprehend problems as diagrammatically consistent throughout the universe.

The demonstration of different methodologies that demand new expressive languages do not only define and quantify what is happening, but express the effect of being exposed or expressed. Institutional opening out, for further discussion on the role of the arts to critically challenge the sciences in terms of their claims about the world, would therefore be required for the benefit of both arts and science disciplines. Perspective formalizes a scene through an imitational scale whereby objects are perceived to recede into an ephemeral apex (perspectivalism) relative only to the gaze. This perspective is relative to where we are and how we think spatially, but with quantum field theory we have a planar example where everything large or small is on the same surface. 'Spacetimematterings' is where things that

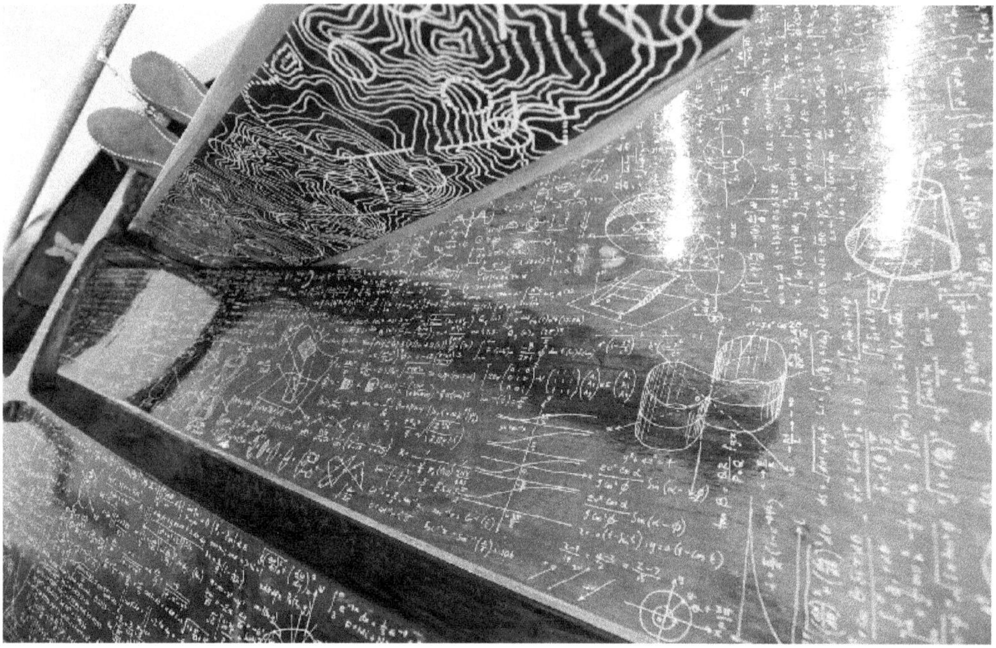

Figure 13: Richard Goodwin, *Moth* (detail). Courtesy of Richard Goodwin.

seem small and things that seem large can become enfolded in one another so that there is a part of the large inside the small, and vice versa.[8] In perspectival space, one object can appear larger than the other by virtue of distancing – but if you were to imagine them both existing as they appear but on the same plane, the object could therefore be larger than the one inside a smaller object.[9] The possibilities of thinking about scale in this manner enables concepts of entanglement to be not only seen as being affected by distance but also that distance, as a classical measurement, does not exist in the same formula as we comprehend it in classical space.

The concept of scale is also explored by Henderson who writes about fourth-dimensional space and the effects that it had on artistic developments in the early twentieth century. Henderson demonstrates the stimulus for *n*-dimensional space with:

> Shadows, mirrors, and virtual images were added to the four-dimensional vocabulary of the artist by Duchamp, whose approach to the subject was unique in this period. If Duchamp at first shared his Cubist colleagues' idealist belief in the fourth dimension, his attitude quickly became more analytical. For Duchamp the n-dimensional and non-Euclidean geometries were a stimulus to go beyond traditional oil painting to explore the interrelationship of dimensions and even to reexamine the nature of three-dimensional perspective.
>
> (Henderson 2013: 205)

Henderson explores the work of Cubist, Futurist, Dada and Surrealist artists in terms of the need to reposition the presumed and fixed observer in classical space, as influenced by the developments (and implications) of fourth-dimensional space and Einsteinian relativism in physics.[10] For Boccioni,[11] as for Picasso (although for different reasons),[12] the classical, perspectivalist spatial plane was being unworked in such a fashion that renders Futurist and Cubist imaging as precursors to quantum field theory (visualized by Feynman's diagrams), which was also developing in the 1920s. Relativity thus impacted painting (Laporte 1988), but painting impacted relativity by giving it a new presence and intuitive form of becoming. The ongoing awareness and assimilation of quantum phenomena is seeing artists attempting to work through the ongoing effects brought to the world by science of shifting realities that affect being human.

To try to elucidate the complexity of chance meetings and historical investigations, Picasso was seen to be exploring four-dimensional space or a form of space-time. Meetings and discussion that took place in the literary café society are testament to the questions that were being posed in relationship to the scientific discoveries that were making waves.[13] This connectivity with other disciplines was extended by the Futurists, whose ideas can be traced to Bergson's philosophical ramifications for defining space and time. For instance, Brian Petrie writes in *Boccioni and Bergson* that 'Poincaré's widely read scientific writings show various affinities with Bergson, and even were this not so they would, as not too specialized sources of information on a wide range of current scientific positions, have recommended themselves to the Futurists' (Petrie 1974: 144).

Entanglement and space

Information encoded in a quantum state is in correlation to its entanglement. Each entangled qubit holds information that can only be relevant to the whole. It is suggested that if you read ten pages of a book of one hundred pages, you will understand a tenth of the whole book. If you have ten pages in a quantum state, you could not understand anything due to the information being in a state of entanglement (Smith 2013). This is because the information in quantum states collapses when they are observed or measured, and thus become a single state. This means that information that is made visible when measured needs to communicate the rest of that information to the other entangled qubit instantly. This information is transmitted faster than the speed of light.[14] Musing on this idea, Karen Barad suggests that:

> Questions of the nature of measurement – or, more broadly, intra-actions – are at the core of quantum physics. Intra-actions are practices of making a difference, of cutting together-apart, entangling-differentiating (one move) in the making of phenomena. Phenomena – entanglements of matter/ing across spacetimes – are not in the world but of the world.
>
> (Barad 2012c: 647)

The 'im/possibilities' of things coming into existence echo the murmurings in the vacuum energy where the slightest perturbance can lead to the formation of a new cosmos.[15] The discovery of the Higgs boson was a monumental moment in human history and in the development of our comprehension of a fundamental core of our existence within the universe. The amazing human awareness and investment into those experiments, permitted through the technical permissions of the Large Hadron Collider, demonstrate a commitment to discovering and comprehending a creative impulse and universality. The materiality of the universe is directly related to our own atomic construction that we are reflecting upon. We observe, as such, and seek to unlock the secrets of the atomic-quantum world, a world of which we are co-constituted, of which we are physically constructed.

A cybernetics of the third kind is where we align our own 'atomic computer' (the brain) with the quantum-atomic universe. This is what is at stake in quantum-atomic analysis and creative introspection. This connection is indeed a phenomenon to be, and has been, honoured. We are the embodiment of our own question reflected back to us by the very nature of the cerebral enquiry (the quantum-atomic matter of the question). To describe the characteristics of any phenomena is a complex process, but by observing various iterations of a given phenomenon and how it functions, we may start to perceive a pattern. The question of patterns is to see them as co-constituted parts of a larger problem. The Large Hadron Collider (LHC) experiment that led to the discovery of the Higgs boson was a self-conscious, self-reflective and focused experiment that pre-empted the outcome, and is also prescriptive in its affect. The LHC experiments defined a singularity that needed to be expressed in context to the whole: the existence of bosons.

A complementary (albeit much earlier) experiment to those conducted through the LHC was undertaken 500 years earlier in Italy by Renaissance artist Carlo Crivelli, who made a career out of imaging biblical events. The experiment to represent a spirit is particularly evident in *The Annunciation, with Saint Emidius* (1486), where a sudden transmission of light expresses visually the inception of a new life in the world. Crivelli, confronted by a need to create an expression of invisible phenomena, a visual vitalism, attempted to create a diagrammatic image of the unknown (immaculate conception). The diagrammatic form of human expression at the core of the unknown universe and becoming can be seen as a vehicle for physicists to develop, in support of the expression of probabilities created in physics. Visualizing ideas as painted diagrams, the links between physics and the manifestations of art, is the sketch that reveals subliminal meanings. It is these meanings that transform and stimulate the work of art enabling invisible links to be forged.

Multiverse

The narrative for this chapter began with the development of the media art installation *Atomism*, which evolved into its latter iteration, *Multiverse*. The research for the initial project *Atomism* led to Feynman's 1979 lectures, where his diagrammatic blackboard presentation visualized the probability of photons reflecting from the surface of a mirror.[16] The concept for *Atomism* was cited in my previous publication *Nanoart: The Immateriality of Art* and was based on the investigation of silver, mirror surfaces and quantum theories of light to see what the viewer substituted for the loss experienced in observing their reflection from a mirror.[17] In Feynman's diagram he reveals how all points on the surface of the mirror receive and reflect light based on the speculative spin of the photon that is not visible to the viewer. The diagram reveals that photons do not merely encode all the information about an object but are also independent of an observer.

Multiverse, like the development of Feynman's diagrams, grew out of unexpected results from witnessing the visualization of *Atomism*. The visual presence of the work was based on the use of an image captured in real time of the individual's portrait as though they were standing in front of a mirror. Each portrait was represented as being analogous to the spin of one photon at one frequency based on Feynman's 'probability amplitude'.[18] The Feynman diagram demonstrates the photon in question as only having one angle, where their spins are in a similar direction that creates a visible portrait in the mirror. This newly reflected portrait's position is at the shortest angle of incidence where the reflection is made visible to the viewer. The development of *Multiverse* revealed all the alternative quantum worlds made visible via this work that are equally real and exist in probable parallel universes.

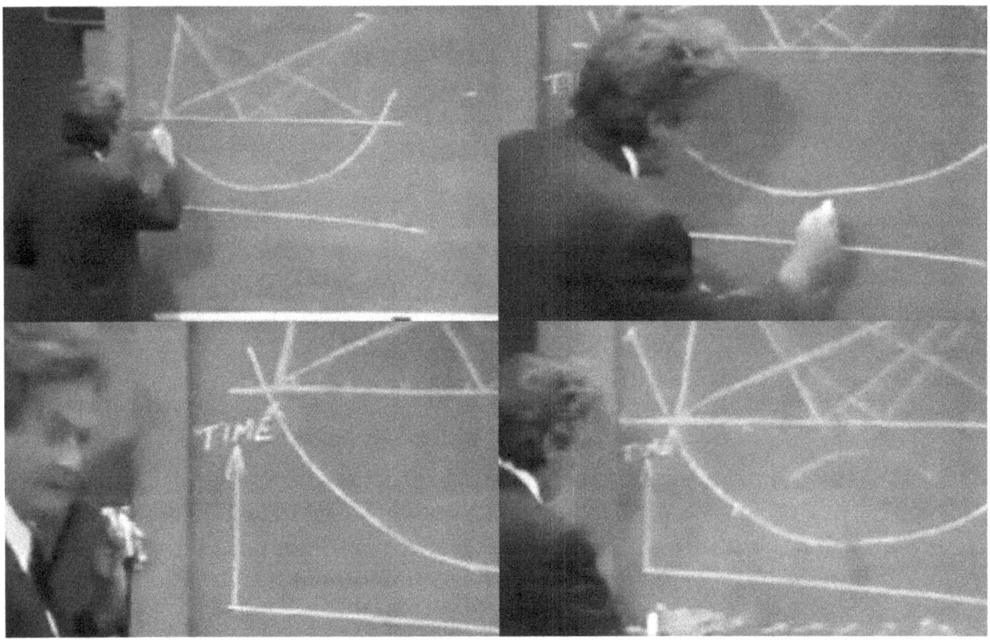

Figure 14: Richard Feynman, *The Douglas Robb Memorial Lectures* (Part 2: Fits of reflection and transmission – Quantum behaviour), 1979. Video (still). Courtesy of the University of Auckland.

In Feynman's lectures, we see him in the act of constructing a diagram so as to image and convey the probability of a photon reflecting from the surface of a mirror (Fig. 14). The gestures and indecisions in his marks are indicative of the complexity of the problem of visualizing thoughts. In one section of the lecture, Feynman tries to use colour to disambiguate the complexity of the problem, where he admits a difficulty, 'let me see if I can do this with colours here; I don't know how to do it with colours, I need to figure it out ahead of time' (1979: 33 mins). Of course, the creative nature of the problem that served to inhibit Feynman's expressive capabilities is clearly identified by Feynman. Here we see the functional role that creativity plays in world-modelling and pedagogical expression and the necessary coalition of art and science.

The complexity of the use of colour reflects on the aesthetic and logistical dilemma experienced by artists in the attempt to construct meaning in the liminal spaces outside the direct expressive capacity of marks and forms (that is, the established visual language). By opening up the palette of complexity – as we see Feynman doing in his lecture – we are seeing it become harder for Feynman, in that moment, despite the ultimate recognition that this expanded palette would only serve to increase his communicative powers. In other words, the live performance of a diagram throws up new and unexpected possibilities that translate to the evolution of the diagram in real-time, where the presenter is under the same pressures of the artist. This problem is the lack of mastery of expressive form, whose availability is ultimately meaningless, if not actively disruptive, in the hands of the technically lacking.

Leonard Susskind bemoans the fading performance of the physicist in front of the blackboard armed only with chalk. He claims new display technologies now only serve to distance the speaker, who previously (with their chalk) would and could, despite the pressure of an audience, attempt to produce a diagram that was embodied in part as pedagogical imaging and live active performance as a pedagogic agent for successful communication of thought. Susskind writes:

> When giving a lecture a physicist would get up with nothing but a bit piece of chalk in his hand to defend himself and to attack you and do whatever he was a going to do. That professor would get up and there was body language there were gestures if he was any good there was ballet sometimes they were bad [...] you know who of course was a master of this was my good friend Dick Feynman.
>
> (Susskind 2013)

We see the drama of the embodied performance happening at the moment of translation from thought to diagram, compounded by the active reception of the audience. The physicist has to therefore perform the diagram in the moment where the probable mistakes and intuitive controls sit on the same surface of the blackboard-cum-void. This performance points to a difference between the quantifiable shareable diagram and what happens in the live presentation of the diagram. There is a mix that demonstrates the point of the observer in the scientific experiment, revealing the problem of human interference with the subject.

Multiverse and parallel worlds

The subject as the diagram was a driving concept behind the development of the art installation *Multiverse*. The work evolved following previous research from imaging of a 7nm (nanometre) silver substrate scanned via the Atomic Force Microscope (AFM). Dr Thomas Becker, a research fellow in the Department of Chemistry, carried out the scanning at Curtin University. The scan is used in the work to act as a filter for a live camera feed image of the viewer's portrait. The scan of the silver substrate qualifies and demonstrates the loss that is experienced when looking in a mirrored surface. The viewer's real-time reflected and analogue portrait is mathematically mapped to the projected spin of a photon screen via the silver filter. In this exposure, the portraits are presented back to the viewer, thus showing all possible representations, not just those closest to the angle of incidence, unbound by the grip of fixed perspective.

The resultant quantum portrait produced by *Multiverse* is generated (for each portrait) via computer, where nearly two thousand real-time digital simulations of the 'sitter's' portrait are consequently indexed by quantum mechanical processes. These multiverse portraits are thus equivalent to Feynman's 'summing up' of the angles of probability. The computational algorithms allow for the processing of these portraits as a direct reaction to the 'probability amplitude', which is represented as an arrow in Feynman's diagram of the photon's direction (Fig. 14). It is the atomic spin of a photon touching the surface of the screen that functions metaphorically as a quantum mirror. *Multiverse* creates a visual diagram that conceptually corresponds to Feynman's but endeavours to take it beyond the standard format of the diagram and extend it into the realm of aesthetics and art.

Multiverse is a response to Feynman's hesitation before the daunting and latent power of coloured chalk. The desire to go further into the realm of art that Feynman was so interested in remains purely figurative, seemingly not breaking the bonds of representation to confront abstraction. The diagram translated and transformed into a work of art has the potential to reveal a different language for what is being described.

Multiverse maintains a strong historical and contextual relationship to the photography of Étienne-Jules Marey.[19] Marey's research work produced photographic and machinic diagrams to define and analyse unseen aspects of objects movement, just as *Multiverse* does for photons. As a scientist, Marey used the camera to make the invisibility of movement visible to the naked eye.[20] The artwork *Multiverse* reveals (indirectly) the invisible spin of the photon that manifests as layered quantum portraits. On the surface of *Multiverse's analogous* mirror, due to the mathematics informing the work, there are points where there is clarity in the coalescence of reflection. For example, the centre image of Fig. 16 shows a spinning portrait that nevertheless appears representative of the viewer. The portrait that *Multiverse* produced in this instance not only exists at the shortest angle of incidence but also unexpectedly reappears at the peripheries of the mirror's surface (left and right panels, Fig. 16).

Multiverse is designed so that the viewer perceives the work on entering a gallery space, wherein three reflective LCD screens are mounted. As the viewer approaches and positions

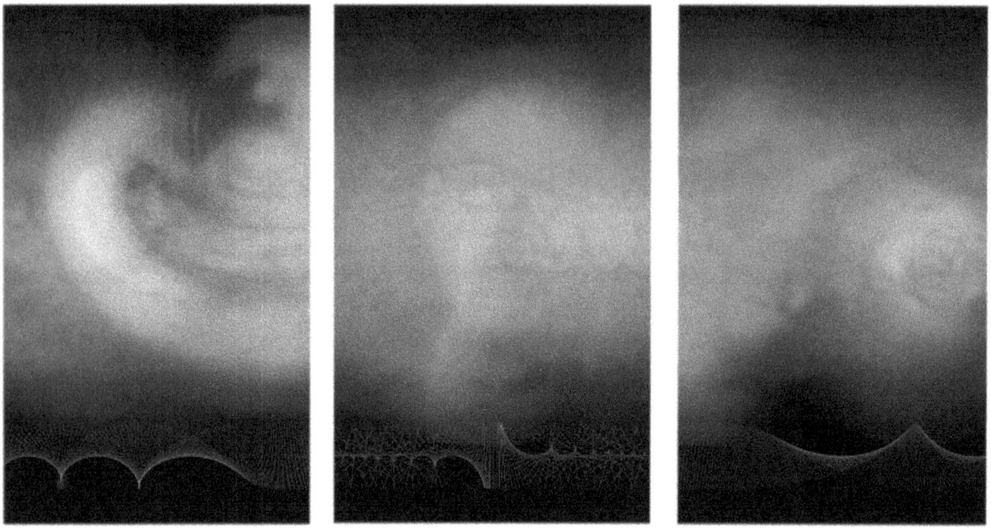

Figure 15: Paul Thomas and Kevin Raxworthy, *Multiverse*, 2009. Screen capture of interface controls showing the AFM scan of silver being used as a filter over the 'sitter's' portrait. Courtesy of artists Paul Thomas and Kevin Raxworthy.

themselves within the spotlight, a camera captures their real-time portrait that is presented in the centre LCD screen. Immediately in the screens on the left and right side of the centre, selected viewpoints of possible parallel universe images of the viewer's portrait are presented. These parallel portraits – the outcome of extracting the data from the Feynman diagram – are revealed via the positioning of the spinning lines (arrows) at the bottom of the image (Figs. 15 and 16).

The concepts of parallel universes and multiverse were originally formulated by Hugh Everett III in 1957.[21] Everett, whilst developing his Ph.D. thesis, implied the possibility of the existence of alternative realities branching off, ad infinitum, from causal events in the actual world – an infinite fractioning of the present. In the formulation and visualization of Feynman's probability amplitudes, Everett's many-worlds interpretations were in various ways validated. The main thrust of Everett's premise took the position that the observer in the world of quantum phenomena was situated in the same quantum state as what is observed. Not the classical observer looking and measuring, therefore collapsing the phenomena to one state, but a quantum observer. This quantum observer, whose gaze is quantum, enacts with each decision a separate state to be in one place and therefore splits into two separate states.

These links between Everett's speculations and modern scientific discovery did not end in the 1970s. As *Multiverse* intends to show in its projected re-emergence of the viewer's portrait, Everett's multiverse conceptually reappears, reified by the seemingly paradoxical nature of quantum physics.

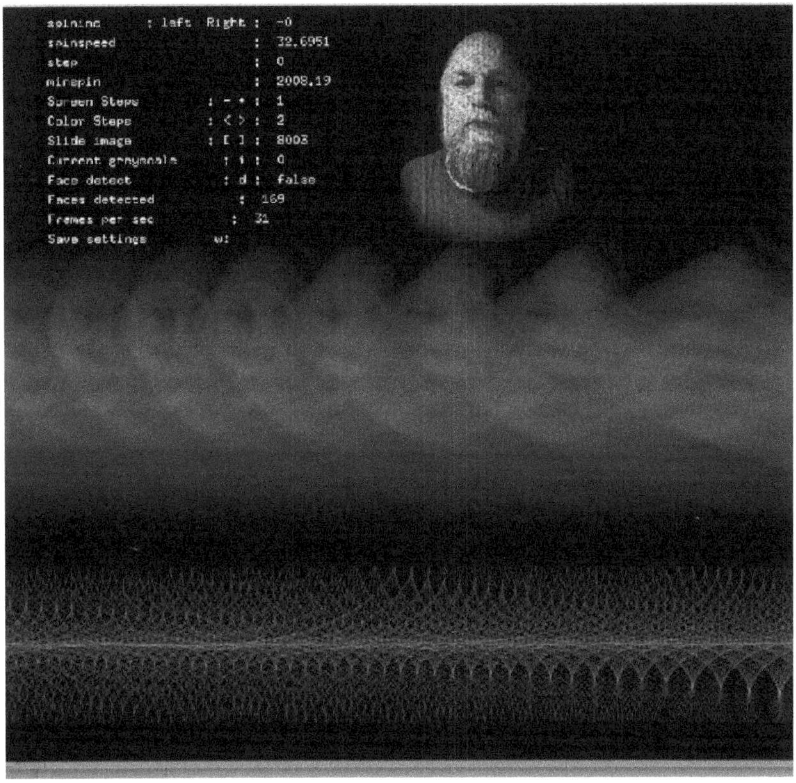

Figure 16: Paul Thomas and Kevin Raxworthy, *Multiverse*, 2010. Installation. Still from the installation with spin lines demonstrating the reappearance of the face in the left and right panels. Courtesy of artists Paul Thomas and Kevin Raxworthy.

Probability of facts

In *Introduction to Metaphysics*, Henri Bergson, discussing the concept of duration, states that:

> An essential characteristic of the concepts and diagrams to which analysis leads is that, while being considered, they remain stationary. I isolate from the totality of interior life that psychical entity which I call a simple sensation.
>
> (Bergson 1949: 43)

In this context, the Feynman diagram can be seen to represent frozen moments in time that want to remain constant (or are held there in cognition). Accordingly, as the diagram changes and evolves, it comes to depict, rather than a unique and fixed thing: 'not a single sensation but several successive sensations' (1949: 44). Bergson suggests that this is 'what science needs for its own proper development' (1949: 44). In other words, scientific inquiry needs to be able to quantify those sensations that are portrayed through the diagram.

Bergson's argument finds its antithesis within the arts, where there is no single sensation that 'does not change every moment since there is no consciousness without memory', to cite his own reckoning of this counterpoint (1949: 11). The ongoing relationship to the world around us as one memory is always 'swelling',[22] in such a fashion that the past mixes with the present into a durational sense of being. The wonder of the diagram is that it can do much more than simply plot the sensation or sensations by adding another level of abstraction that allows for possible paths to be perceived. These possible paths of problem-finding (not problem-solving) must be what Bergson sees as constituting what we commonly call 'intuition'. The diagram, when it is allowed to follow the abstraction of thoughts, provides a platform for constructing new concepts and thus enables the diagrammatic process to reveal and articulate the emergence of latent problems (1949).

Feynman's diagram is positioned in a symbiotic relationship with the aforementioned Deleuzian conception of Bacon's painting practice (Figs. 17 and 18). Deleuze uses the concept of the diagram when he references Bacon, commenting that the act of painting consists of

> random marks (lines-traits); scrub, sweep, or wipe the canvas in order to clear out locales or zones (color-patches); throw the paint, from various angles and at various speeds. Now this act, or these acts, presuppose that there were already figurative givens on the canvas (and in the painter's head), more or less virtual, more or less actual.
>
> (Deleuze 2005: 99–100)

The agony of Bacon is not diminished by the cerebral, but rather is all the more heightened by the causal relationship of the desire to reveal what cannot be seen. The diagram as under-painting evident in Bacon's painting (Fig. 17) is consistent with primordial chaos creating the perturbance from which worlds ultimately are formed. Feynman's diagram attempts to explore the inarticulate means of describing the quantum world. Passion in the diagram is the controlled passion of a spiritual quest that can lead to the rupture of emotion, but this rupture (in Bacon's case) is still inscribed in the palimpsest created from the under-painting.

For Bacon, the process of painting forms the basis for interpretational (ontological) problems to be made real, visualized through acts of smearing, scrubbing, sweeping and wiping (Fig. 18). Therefore, the diagram for the work, and of the work, evolves. Various modes and means of engagement with physical materials reveal things (by virtue of their unique character) that distort the status quo in representation and open up new possibilities in our experiencing of the world. The marks in Bacon's diagram are dealing with the physics of, and are mediated by, the latent physical properties of a material conduit (medium), not in order to solve a problem but rather to expose the probability of something existing within the work. In other words, this creative 'revealing' opens a feedback loop that guides the

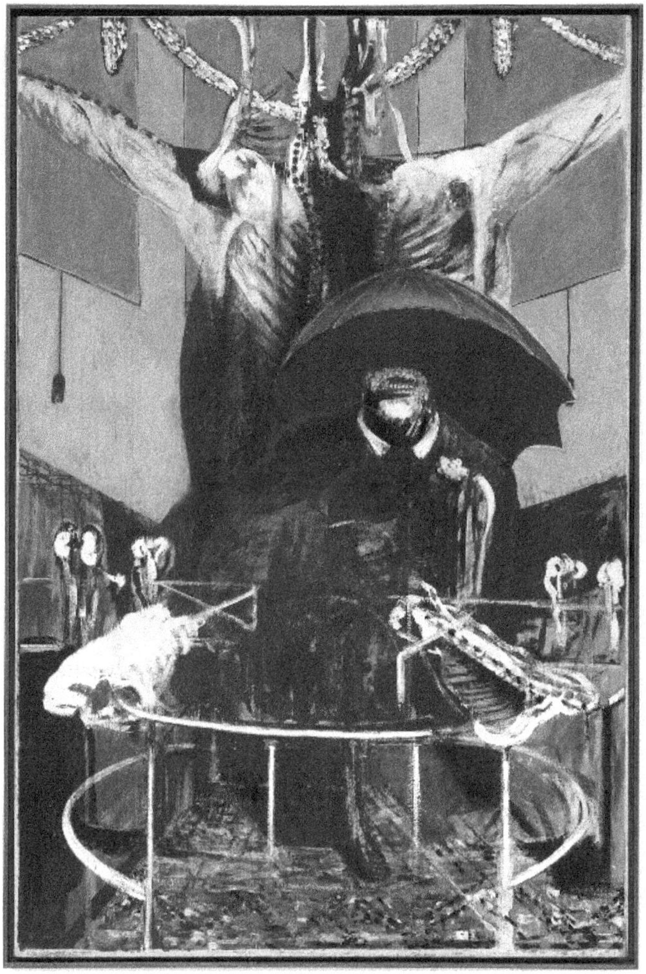

Figure 17: Francis Bacon, *Painting*, 1946. Oil and pastel on linen. 197.8 x 132.1 cm. Courtesy of the Museum of Modern Art (MoMA), New York.

image and gives form to the intuition of its author. It is this feedback loop that is the speed of thought.

In regards to Bacon's diagrammatic painting, Deleuze writes, 'it is as if, in the midst of the figurative and probabilistic givens, a catastrophe overcame the canvas' (2005: 100). Probability is what is being explored and what is revealed in Feynman's and Bacon's diagrammatic marks that proffer their respective and intuitive grasp of the inaudible (but latent) murmuration of the quantum world. This foundational work defines a way of visualizing thinking as though the probable painting was already in his head, and the process of visualizing this creates the fundamental flow, a sense that Deleuze captures in his

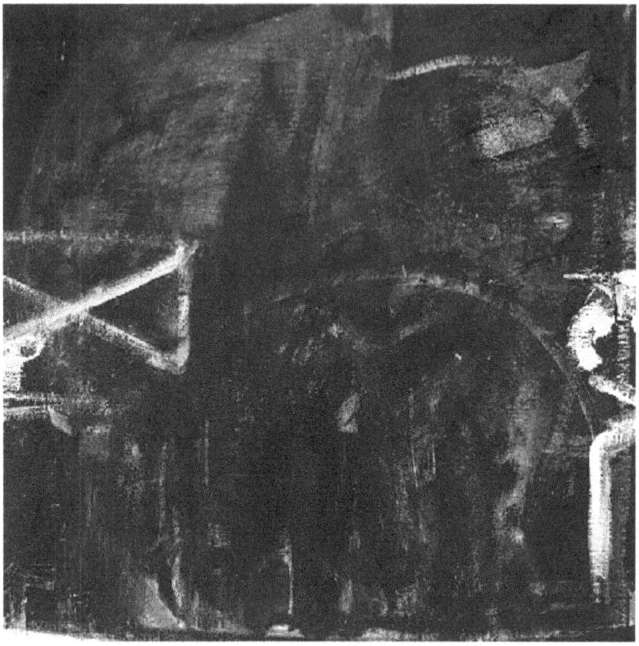

Figure 18: Francis Bacon, *Painting* (detail), 1946. Oil and pastel on linen. 197.8 x 132.1 cm. Courtesy of the Museum of Modern Art (MoMA), New York.

line: 'it is like the emergence of another world' (2005: 100). By fleshing out this world (so to speak), Bacon's meaty but emergent forms resonate with quantum conditions of becoming and emergence. Through Bacon's painterly gestures we are consequently allowed access into the strangeness of the quantum-atomic world.

In aid of the diagram

With the aid of the diagram, 'entire new calculation vistas opened for physicists' (Kaiser 2005a: 4). This enabling of physicists to explore the potential of diagrams that can be used in calculations extended the scope of their significance. The diagram took on a new status, but in allowing for the visualization of probabilities, they were not extended into the more subjective relationships of the diagrammatic existing in fine art.

The scientific diagram can never be the phenomenon; it can only be a series of expressions of moments or states. The dots, arrows and lines that individualize the probable position and direction of matter presuppose its singularity. The problem comes in describing the quantum as a field that is everywhere and taking all possible routes to any given target. Here Feynman's diagrams become expression of a more holistic understanding of the objectification of time and space through the inherent quality of linearity. Bergson's concepts of duration, perception and the subconscious have pointed out to us that time is not the instrument of measurement but a place where past and present coalesce. The duality of time and space is borne out by quantum theory. It enriches Bergson's theory of duration by showing the unmeasurability of a thing if it wants to remain in a fluid state. The definition of the power of a line is seen in its potential to symbolize a state that can be used to formularize quantum phenomena.

Marta de Menezes, writing on Feynman in her master's thesis, addresses why his diagrams

> can be used as a mathematical tool is due to the following characteristics. First, the exchange of a photon represented by a wiggly line should not be taken as a classical particle following a single space-time path, but as the sum of all possible ways in which that photon could have gone from one particle to the other.
>
> (Menezes 2001: 34)

The Feynman diagram (Fig. 15) is infusing the code with stable scientific meaning for the probability amplitude of plotting the path of a single photon's reflection from a mirror's surface, presented as a particle state indicated by an arrow. The diagram is attempting to define the wave-like state as reflected in the terms of the wave–particle duality.[23] The change in state of the wave–particle duality is constant in all atomic structures. When measured, they act as particles and when they are not being measured, they act as waves. The particles in their wave-like state decide on which routes to take in their travels by summing up all the options, in the same way that we might do when confronted with a queue at a shopping market. Neil Turok describes in his CBC Massey Lectures 'The universe within: From quantum to cosmos' how when we decide to join a shopping queue, we look at which one to join by gauging how many people are in the line, how far the queues are from our location, whether there are families, how many goods are in the basket etcetera, and then choose a route.[24] Particles do the same thing, based on William Rowan Hamilton's action principle that suggests particles survey all the possible routes to the future and choose the one that demands the least actions. Hamilton suggested that physical systems could even take a multitude of different routes with probability amplitudes for each part based on each action.

As the complexity of quantum became more immutable, the amount of data that needs to be processed based on probabilities of something occurring will reach the limits of what binary computing can do. In Feynman's case, he changed the rules in science by using visualization as a means of more fully understanding the probability that some phenomenon can happen.[25] In the future it will be the atom itself that records and processes its own data; each qubit will allow for hundreds of calculations to happen simultaneously. How will the diagrams of quantum computing develop to complement and visualize this new future? The likelihood is that artists are simultaneously working intuitively on these visions of how quantum computation might serve as the platform through which to understand quantum mechanics.

Material deconstruction, the breaking down of matter to its atomic fundamental particles, could enable the quantum computer to more readily process quantum information, but carries with it the risk of incestuous connections. The diagrams of science are particularly designed to substantiate and prove the theories. The scientific instruments are duplicitous in this desire to substantiate the theories and therefore prove their own existence. Controlling the materiality of what constitutes nature enables us to process data pertaining to the very nature we are seeking to comprehend.

Photonic traces

Feynman's description of photons referenced in his 1979 presentation has their spin at 10^{-15} (quadrillion) that prompted a concept to explore how much data a photon collects on its journeys, and of course what traces of electromagnetic activate does it leave when it hits a surface structure. The photon itself carries information that is embedded in its structure. Therefore, a swarm of photons in turn carries the mimetic images of things they are affecting and being affected by as they move through space and time. Could this space, when filled with a thousand million visual objects all moving through one another in a formless mass, be recorded or understood? If they were to be understood, what form would the sum total of everything look like? Did the artists of the Impressionist period understand the very nature of questioning light at its fundamental level of particles (and waves) by trying to capture in their paintings not the object revealed by light, but the behaviour of light that particles (waves) exhibit when interacting with objects?

The photons are archiving the visible world independently of it being observed by any other thing. For example, the redshift happens when light is transmitted from an object such as a planet in an outer galaxy. In this context, the wavelength is stretched due to the expansion of the universe, revealing a red spectrum that can enable cosmologists to measure distances that the outer galaxies are away from earth. The measurement of the light coming from the galaxies enables information to be gathered about distance, but what if we could get more data from the photons travelling through space? The last object that the photons struck, where the photons expelled from, could still have part of the trace energy of its atomic characteristic. Is there a potential for reverse engineering of a surface's accretion of electromagnetic activity that could reveal in the strata its history of interaction with other physical materials?

This would mean that the world is being archived constantly in its quantum immateriality by being absorbed into matter. The palimpsest inscriptions found in the over writing of text upon text would possibly not produce a linear readable output. The layering of data would demand a different understanding of what constitutes linearity, time and space. Matter could record more information of the world by the potential of photon's energies to stimulate and excite the material substrate it contacts. As with the reflection into the silver of a mirror, we understand there is a loss, as some photons are absorbed and some photons are expelled from the surface. The energy from the photon that is absorbed is a potential archival force that transmits its signal, which is waiting to be decoded and reengineered to form a new meaning and context within the matter itself; to unlock the data of its history. This is the idea that a movie plays on a screen and at the end of the movie the sum total of the movie is recorded in the materiality of the screen.

Bergson's diagram and quantum parallels

Bergson's theory of duration parallels that of the wave–particle duality postulated in the beginning of the twentieth century in quantum theory. It is a fundamental pre-emption of Werner Heisenberg's uncertainty principle. The famous 1927 paper on the topic[26] gave us the popularized beginnings of quantum theory. In the title of the paper, the role of intuition is thrown into the mix of uncertainty.[27] If we were to take the findings that led to our current understanding of the wave–particle duality, it is that a particle appears when it is observed/measured and at all other times it acts as a wave.

The connection between the theories lies in the inability of duration to be seen as being anything more than a state of consciousness. Bergson took the view that we have on the 'one hand a multiplicity of successive states of consciousness and on the other a unity which binds them together' (1949). The clustering of thoughts that become objectifications of time are the realizations of ideas that coexist with the present.

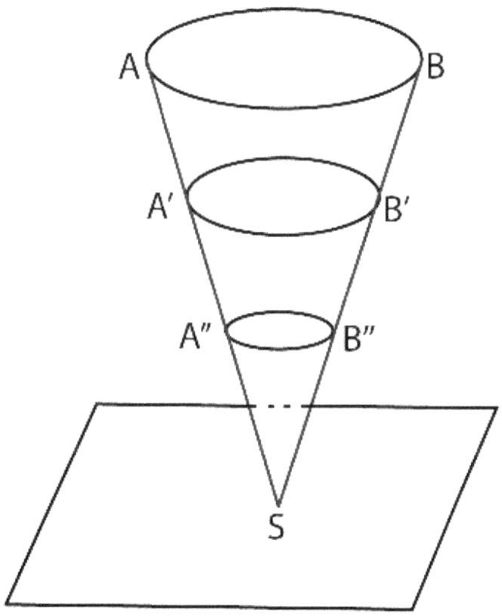

Figure 19: Henri Bergson, Diagram 1896 from the book *Matter and Memory*, 1990.

Bergson developed a diagram that marked out through the use of an inverted cone, a mixing of memory, time and space, down to a final present point (Fig. 19). In Bergson's diagram, the plane is referred to by Simon O'Sullivan as a 'plane of matter',[28] located below a series of conical concentric circles, like waves moving out in water from the dropping of a stone. Bergson was an advocate of wave theory and its effect on all things as 'life appears in its entirety as an immense wave which, starting from the centre, spreads outwards, and which on almost the whole of its circumference is stopped and converted into oscillations' (1911: 266).

The point(s) is analogous to the pencil mark or the cantilever point of the AFM (Fig. 20) in its ability to represent a present as part of the ongoing fluidity of movement. The diagrammatic cone can be seen as moving across the plane of action as well as moving between and through the A&B's, focusing on the point not as a fixed entity but as a haptic barometer on the material world.

In his Ph.D. thesis of 1924, French physicist Louis De Broglie postulated the wave nature of electrons and suggested that all matter has wave properties. De Broglie made a significant contribution to quantum theory and later in his life wrote a revealing essay on the basic similarities between Bergson's insights and the standpoint of quantum mechanics. Bergson's intuitive insights, dating back to his first book *Time and Free Will* based on his doctoral thesis,[29] demonstrates an awareness of the immeasurability of anything within the universe predating Heisenberg and Bohr by 40 years. Lewis Feuer states that 'Bergson had indeed been the intuitive prophet of quantum mechanics according to Louis de Broglie' (1974: 220). Broglie states that the problem that physics describes is the complexity of the uncertainty predictions: 'Does not this quite exact summation of the present position of quantum physics suggest a certain aura of Bergsonism?' (1969: 58).

Bacon's synergies

In this context, I want to draw a transversal relationship between Feynman's diagrams and the diagrams (paintings) of Bacon, as theorized by Deleuze in the book *Francis Bacon: The Logic of Sensation*, and the media artwork *Multiverse*.

In Bacon's well-documented personal library, there was a publication of *Phenomena of Materialisation* (1923). In the introduction, there is a passage that relates to the

> abandonment of the materialistic conception of the universe which, even thirty years ago, was in sole possession. Modern physics regards matter as a form of motion, and is dominated by the idea of energy. Psychology also is gradually emancipating itself from the purely physiological conception of mental life; and under the leadership of the philosopher Bergson, it tends to acknowledge the superiority of the psychical over the physical. Thus the circumstances are much more favourable to the investigation of great new problems and facts than they were some decades ago.
>
> (Schrenck-Notzing 1923: 2)

The reference reveals the complexity that aligns with our current understanding of matter, a re-return to the materiality of thought. Despite this passage suggesting the emancipation of thought from matter or biology (as it did when it gained independence as a science from physiology), now we see in physics the return of the psychical and its materiality; thought may now be seen inscribed in biological and chemical processes that are observable. Bacon's diagrammatic pre-paintings resonate with the coalescence of consciousness and materiality. Max Tegmark suggests that 'consciousness can be understood as yet another state of matter. Just as there are many types of liquids, there are many types of consciousness' (2015: 3).

The particular reading of the diagram in the service of art creates a difference of intention from the scientific towards the 'intuitive'. This 'survival of the past' into the present moment means that we are not trapped in 'instantaneity' but can unfold and allow the concept of a durational 'swelling' of consciousness (Bergson 1912: 12). Within this concept of duration, we formulate a 'perceptual becoming' that allows the diagram to oscillate between different states. Out of these different states, the diagram creates a potentiality of outcomes that can

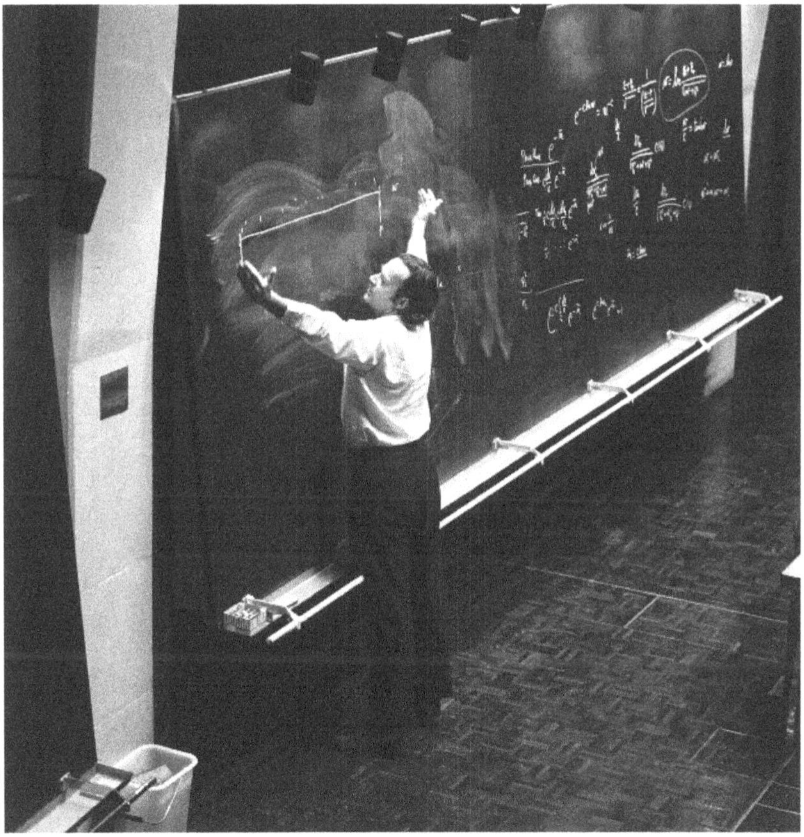

Figure 20: Richard Feynman lecturing at CERN. Courtesy of the CERN/science photo library.

be visual but not visualized. In other words, the abstract quality of Feynman's diagram is that it expresses so much to the physicist, in that it identifies points in time as frozen moments and as a summing of all possible futures. These dualities can be seen as the conceptual basis for the creation of artworks, but the works of Bacon are extending the diagrammatic nature of comprehending life. The artwork attempts to create human interaction with the possibility of becoming and being in the world. The artwork responds to the various inputs and stimuli but is indicative of a state of becoming that can only be viewed in its semi-abstract state as a conscious stream. De Menezes states that in relationship to the artist Joseph Beuys' diagrams, they form 'a self-consistent world, where some personal conventions are used [...] to identify features in the blackboards, their meaning and contribution for the global sense he was trying to transmit' (2001: 43).

The blackboard (the scientist's canvas) as a void, ether, nothingness, a space to create a sense, an articulation of something new happening through the performative diagram, is going straight to a state of exploring becoming – an abstract state. The diagram exists as a palimpsest; the smears and scrubs made by erasure are partly disguised by the previous marks, which are visible in Fig. 20. Feynman's performative gesture, corresponding to the line of measurement on the blackboard, explains an embodied agency within the marks being made.

Deleuze suggests that Bacon uses language similar to Wittgenstein,[30] invoking the diagram to explore the 'possibilities of fact'.[31] It is this very possibility of the painting that ties in with Feynman's diagrams. The 'possibilities of facts' are not in themselves facts but probabilities of all possible states. The diagram that demonstrates how the spin of photons reflects from a surface uses the probability amplitude to perform a process of discovery. When creating the diagrams on the blackboard, Feynman wipes away, moves between, and so forth, to lay out the diagrammatic foundation of presenting the probability of an event. What Bacon does by producing the visualization from the probabilities revealed in the smearing, scrubbing and rubbing is a post-diagrammatic presence in the work exposed for all its flaws and possibilities. The faces in some of Bacon's triptychs, with their smears and voids, give us a view of the world never seen 'more or less virtual, more or less actual' (2005: 86).

The readings of the two forms of constructing the diagram are to show the symbiotic and complex relationships between science and art. While choosing the two men seems counter-intuitive, both Feynman and Bacon were striving for the oscillation between probability and possibility that something was happening in the world, emerging from humankind's continual probing and imagination in attempting to comprehend the universe. The diagram forms the platform for the potential of a new visual language to emerge, not as pure measurement but by activating the surface upon which it is transcribed. Can the diagram, as a projection of thought, sketch out or map paths that lead to some remembered destination that is hidden within the mind (intuition)? The diagram draws out from a subconscious mental architecture to translate remixed memories into some form. When you try to visually describe what is in your head to explain something to somebody, the diagram begins as an onomatopoeia, a performative language where there is potential for a sonic description. In the case of a map that lets someone know directions, the diagram is instructional; it takes the form of clarity that demands a mental spatial acuity. The adjustments that need to be made interfere with the acuity of projection when shifting spatial scales of thought to adjust to actual physical boundaries.

The subject of the mirror in *Multiverse* is an investigation of light that puts the viewer's reflection at the shortest angle of incidence but also at the point of collapse where the viewer becomes the subject that is seen: the object and the subject combined. Maurice Merleau-Ponty suggests: 'Thus since the seer is caught up in what he sees, it is still himself he sees: there is a fundamental narcissism of all vision' (1968: 139). This is not a question of cognition but rather a question of quantum material and concept interaction. The photons carry the experience captured from the absorption and emission, acceptance and resistance of energy to and within an object's surface. The emission is then transmitted to the next atomic structure where the material accepts that encoded transmission, to be decoded and retransmitted. The illusion of the perspectival paradigm is secured when the 'painter' cannot separate themselves and mistakes the scope of the mirroring they are seeing for the total picture.

Artists explore the world through a reflective internalizing comprehension of phenomena. It is this that creates the strange syncretic relationship with scientific discovery (shared origin of both in the diagram). The world of discovery is the cornerstone of practice and research that draws connections to facilitate a change in human perceptions and understandings of being part of this universe. The quantum phenomena of Feynman's diagrams are inherent in the marks, rubs and scrubs of Bacon. The contemporary convergence of art and science is formed through the potential of the diagram to speak with one shared voice as a purveyor at the intersection between gestural expression and logical expression.

Notes

1. *Bull* (1945) provides a famous series of eleven lithographs that demonstrate Picasso's personal vision of how to create an abstract artwork from a representational drawing of a bull.
2. The planetary model of the atom was introduced by Niels Bohr in 1913. In this model, the electrons move around the nucleus as the planets orbiting the sun. Unlike the planets, the electrons are not subject to gravitational forces but respond to discrete values. This model becomes translated into the common three-dimensional image we have come to universally recognize as being a visual and actual representation of the atom.
3. Paul Klee's book *Pedagogical Sketchbook* was 'intended as the basis for the course in design theory at the famous Bauhaus art school in Germany'. The book demonstrates the interpretive nature of line work and its ability to express subtle, complex languages (Klee 1961, 1968).

4 Kandinsky's painting *Several Circles* (1926) relates to notes from his book produced while working at the Bauhaus. Kandinsky, in his 1926 textbook *Point and Line to Plane*, described the point: 'Abstractly or imaginatively, the point is thought of as ideally small, ideally round. In actuality, it is an ideally small circle [...] Here no boundaries can be fixed and the realm of the point is unlimited' (Kandinsky 1979: 30–31).

5 Klee describes three series of different sequences of line as being '1) An active line on a walk, moving freely, without goal. 2) A walk for a walk's sake. The mobility agent, is a point, shifting its position forward. An active line, limited in its movement by fixed points. 3) A medial line which is both: point progression and planar effect' (Klee 1961: 16–17).

6 'Artists are the antennae of the race, but the bullet-headed many will never learn to trust their great artists. If it is the business of the artist to make humanity aware of itself; here the thing was done, the pages of diagnosis. The multitude of wearisome fools will not learn their right hand from their left or seek out a meaning' (Pound and Fenollosa 1967).

7 The Russian Suprematist Kazimir Malevich painted *Black Square* over an existing canvas. The work speaks to a rupture needed to create a visual language that is aware of, and challenges, its time. Malevich's ideas of colour were heavily influenced by the experimental research of James Clerk Maxwell translated into Russian at the turn of the twentieth century.

8 Barad explains what spacetimemattering means by 'putting the point another way, phenomena are patterns of "mattering" – diffraction patterns dispersed across differently entangled spaces and times, or rather spacetimematterings' (Barad 2012a: 77).

9 Doctor Who (Tom Baker) is being questioned by Leela (Louise Jameson) in Chris Boucher's 'The Robots of Death' (1977) to illustrate a phenomenon of why the TARDIS is bigger on the inside than the outside. Explaining that 'it's because insides and outsides are not in the same dimension', the Doctor takes two boxes, one larger than the other, and places the large box on one side of the TARDIS and holds the other small one close to Leela, where it appears larger. The doctor states 'if you could keep that exactly that distance away and have it here, the large one would fit inside the small one'. The Doctor demonstrates what he calls 'trans-dimensional engineering' through the illusion of perspective. This appears to be another definition of Barad's 'spacetimemattering'.

10 The Cubist movement that started in Paris in 1907 moved away from the representation of perspectival space towards a semi-abstract remodelling of the world. Two years later in Italy, the Futurist movement began its transformation of spatiality and the coalescence of matter. *The Futurist Manifesto* by Filippo Tommaso Marinetti was originally published in French in *Le Figaro* in Paris on 20 February 1909 (Apollonio 1973: 19). This was followed by the total reconfiguration of perspective by Suprematism, an art movement from Russia that began in 1913. Malevich's painting *Black Square* (1915) was seen as spearheading the reconfiguration of art. In Switzerland in 1916, the Dadaists began to define their ideals for shifting what constituted being in the world. In 1920, the Surrealist movement was founded, merging conscious with unconscious reality into a 'super reality'.

11 In Boccioni's work, we see the beginnings of a conscious visualization in art of matter as flux, not as a thing. In his art, Boccioni was trying to intuitively understand the scientific and philosophical nature of matter, which would not be made truly visible until the invention of the AFM. The link between Boccioni and Bergson both coming to terms with what is now

current thinking around the permeability of matter is significant here. Boccioni's vision of matter as being part of an 'indivisible whole' correlates with Bergson's durational aspects of rhythm, tempo, vibration and pattern, and draws parallels with the continual vibration of matter at its atomic. The context and influence of Boccioni's work is referenced in my first book on nanoart (Thomas 2013).

12 Arthur I. Miller points out '[h]ow the 20th century's most important scientist – Albert Einstein – and its most important artist – Pablo Picasso – made their greatest discoveries at almost the same time is a remarkable story: Einstein's relativity theory in 1905 and Picasso's *Les Demoiselles d'Avignon* two years later' (Miller 2004: 484). Miller does not state that Picasso or Einstein ever met but suggests that in the small circle of intellectuals in Paris, new scientific concepts were being discussed in cafés and clubs.

13 Miller points out that Maurice Princet, the French mathematician, 'was seen with Picasso's group in cafés, took part in their hashish sessions, and visited Picasso's atelier at critical times when Picasso had problems with *Les Demoiselles d'Avignon*. La bande à Picasso were enthralled by Princet's informal lectures in cafés and bistros on Poincaré's famous book *La Science et l'hypothèse*, all the rage in Paris' (Miller 2004: 488).

14 The concept of quantum entanglement will be discussed in more detail in Chapter 3 (Spin).

15 Vacuum energy is all the energy within the universe. Dark energy constitutes 75% of the universe, and dark matter 21%, about which little is known. They have their roots in vacuum energy, leaving normal matter that we do know about to be only 4%.

16 Richard Feynman, *Science Videos: The Douglas Robb Memorial Lectures* ('Part 2: Fits of Reflection and Transmission – Quantum Behaviour') displays the diagram used in the work of *Multiverse* (Trust 1979).

17 The initial interest in silver was based on its properties to act as the reflective surface of a mirror. Silver played a role in the birth of perspective and the objectification of the world through Renaissance perspective. Filippo Brunelleschi used polished silver in his 1425 peephole device to replace the sky that he painted outside the Duomo in Florence of the baptistery (Thomas 2011).

18 Feynman's lectures, volume 111 on probability amplitudes, states: 'We ask for the probability that a particle will be found at x. Our first general principle in quantum mechanics is that the probability that a particle will arrive at x, when let out at the source s, can be represented quantitatively by the absolute square of a complex number called a probability amplitude – in this case, the "amplitude that a particle from s will arrive at x." We will use such amplitudes so frequently that we will use a shorthand notation – invented by Dirac and generally used in quantum mechanics – to represent this idea' (Feynman 1963).

19 Étienne-Jules Marey (1830–1904) was a scientist working initially in the area of medical devices and later in the realm of image science. He revealed his unique role by using photography to scientifically demonstrate unseen aspects of the material world (Marey 1895).

20 Marey's photographic work can be seen as having great influence on the planar and sculptural practices of Cubist and Futurist artists. The ability to make visible movement that was too quick for the eye to record gave artists a photomechanical way of seeing that, until that moment in time, had been unknowable (Dagognet 1992: 148–49).

21 Hugh Everett's first dissertation was called 'Wave Mechanics without Probability,' with the shorter, highly edited submitted dissertation retitled as 'Relative State: Formulation of Quantum Mechanics' (Everett 1957). The dissertation was not well received, even though his supervisor and mentor John Wheeler wrote a positive review of it. Wheeler set up a meeting between Everett and Niels Bohr to see how he would deal with this new theory. In 1959, Everett went to Copenhagen to meet with Bohr, but he would not entertain any new concepts such as many-world interpretations (Everett 1973).

22 In Bergson, swelling is explained as 'pure duration that we then plunge back, a duration in which the past, always moving on, is swelling unceasingly with a present that is absolutely new. But, at the same time, we feel the spring of our will strained to its utmost limit' (Bergson 1911: 200).

23 The Dual Slit (Thomas Young) experiment is explored by shining a laser through two parallel slits on to a sensor screen that creates an interference pattern (see Fig. 32a). Then, when analysing photons movement, it was found that when the light was dimmed so only one photon could be detected at a time, the interference pattern remained. A photon would pass through both slits acting as a wave and a particle at the same time. Each individual photon interferes with itself. Photons behave like waves and explore all available roots to a source, but when measured they act as particles.

24 In Neil Turok's Massey lectures, he suggests how Hamilton's theories were formative in understanding the quantum of action (Turok 2012).

25 Richard Feynman demonstrated his diagrams in 1947 where he 'told his fellow theorists that his diagrams offered new promise for helping them march through the thickets of QED calculations [...] By the end of the 1960s, some physicists even used versions of Feynman's line drawings for calculations in gravitational physics' (Kaiser 2005b: 158).

26 Heisenberg released his uncertainty paper entitled 'Über den anschaulichen Inhalt der quantentheoretischen Kinematik und Mechanik' (1927). A (partial) translation of this title by Miller states: 'On the intuitive (*anschaulich*) content of quantum theoretical kinematics and mechanics' (1927), where he (Heisenberg) demarcated boldly the notion of how 'we believe that we understand a theory intuitively (*anschaulich*) from the visualization of atomic processes' (Miller 1986: 151). Here, a particularly notable translation of *anschaulich* is 'intuitive'.

27 Miller adds to the concept of Heisenberg's use of intuition, '[t]he importance of the concept of intuition to Heisenberg is indicated by its inclusion into the title of this classic paper in the history of ideas. By redefining the concept of "intuition" he found the new "intuitive interpretation of the various phenomena" for which he had searched in his 1926 paper "Quantum Mechanics"' (Miller 1994: 13).

28 O'Sullivan describes how our 'plane of matter – our world as it were – carries on indefinitely: there are always further objects behind the present ones'. There is a coalescence on the plane of all matter (O'Sullivan 2013: 170).

29 *Time and Free Will: An Essay on the Immediate Data of Consciousness* was translated into English in 1910 (Bergson 1910).

30 In Ludwig Wittgenstein's *Tractatus Logico-Philosophicus*; '2. What is the case – a fact – is the existence of atomic facts. 2.01 An atomic fact is a combination of objects (entities, things).

This epistemic notion is further clarified by a discussion of objects or things as metaphysical substances. 2.014 Objects contain the possibility of all states of affairs. 2.0141 The possibility of its occurrence in atomic facts is the form of the object' (Wittgenstein 1961).

31 In the quote from David Sylvester's interview with Bacon, he states: 'No, the marks are made, and you survey the thing like you would a sort of graph. And you see within this graph the possibilities of all types of fact being planted. This is a difficult thing; I'm expressing it badly. But you see, for instance, if you think of a portrait, you maybe at one time have put the mouth somewhere, but you suddenly see through this graph that the mouth could go right across the face' (Sylvester 1987). Deleuze paraphrases Bacon by using the 'possibilities of fact' (Deleuze 2005: 101).

Chapter 3

Spin

Quantum spin

Over the last century, quantum theory has progressed from a quirky esoteric science to an essential pillar of contemporary physics. The quantum model of the universe is one that shares many conceptual similarities with earlier theories of Atomism – in which the world is constructed from various combinations of fundamental and indivisible units of matter. However, in reality, the quantum and the atomic are not easy partners. When Einsteinian science made possible the splitting of atoms, conceptually speaking, the quest for deeper-than-atomic foundations of the physical universe began – questioning upon what grounds are atoms built? The world that has since been unveiled, the quantum, is a strange and in many ways irrational space. 'Irrational' insofar as quantum mechanics defy the protocols of common sense. At the turn of twentieth century, they immersed the world in uncertainty as being at the heart of what is now becoming our realities. Realities that we cognizant humans have developed after thousands of years of intelligent observation of the world around us; the chemistry and physics of our everyday existence that we intuited before chemistry and physics were developed into tools of analysis and understanding.

In quantum mechanics – the codified activity of quantum phenomena – the word 'spin', in benign historical misfortune, exemplifies the counter-intuitive nature of the quantum world. In quantum theory (and particle physics) 'spin' refers to an 'intrinsic form of angular momentum' that is expressed in both elementary and composite particles such as electrons and atomic nuclei. This is to say that particle spin refers to one category of movement that is quantifiable in quantum physics: one of two categories of 'angular momentum' in the field, the other being 'orbital angular momentum'. This second kind of 'orbital' angular momentum is the quantum equivalent of orbital revolution in classical physics, of massive bodies dancing around each other in the cosmos, like the moon around the Earth and the Earth around the greater mass of the sun. Spin, on the other hand, is comparable to what is called a 'vector quantity' in geometry, meaning that the very measurement of quantum spin serves to designate a definite sense of magnitude as well as a sense of 'direction'.

Given the extraordinary difficulty of measurement at the subatomic level, it might be sensible to qualify 'spin' – as this term is central for the discussion that follows – as fluctuation that cannot be accounted for in terms of a product of angular orbital momentum (Fig. 21), the other category of angular momentum discussed above. These analogies all have problems in trying to describe quantum phenomena within classical values. Orbital angular

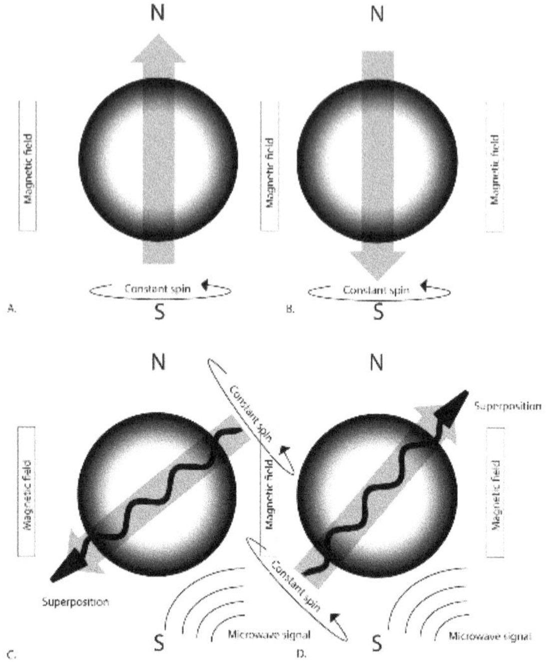

Figure 21: Illustration of electron spin between North and South. In (C + D), the microwave signal generates an orbital angular momentum creating a probability of a position between north and south. Graphic by Paul Thomas, 2018.

momentum defines the shape of the wave function of the electron around the nucleus, but the delocalized electron cannot be defined as orbiting the nucleus. When we apply our macro-sensibilities to this problem, we might be confused as to why the terminology is so unapproachable if the behaviour it seems to describe is so natively intuitive (referring again to the orbit of celestial bodies bound by the mechanics of gravity). This confusion should be dispelled (or rather, replaced with a more apt and existential confusion) when we remember that no atom, let alone subatomic particle, has ever been seen or can ever be 'seen' in the sense that we commonly understand – that is, with our biological eyes. If we recall that these bits of matter in question are bound to the same physical rules as the medium through which we see, visible light, then common sense goes out the window.

The quantum spin of a given subatomic particle is an intrinsic property that categorizes the magnetic behaviour of particles. 'Intrinsic' because it is the empirical spin of a particle that gives it categorical identity. This is to say that the spin momentum of a particle is deterministic but, unlike categories in biology for example, it is shared exclusively and absolutely by particles of a given category; unique only as categories, not individuals within a loose taxonomic category. Each given kind of particle is designated with a given spin number determined by 'intrinsic' angular momentum. Using the analogy of rotation, the rate would not be fixed; the rate depends on the external magnetic field affecting it. With Quantum mechanics we are aware these are just analogies; the particles are not actually spinning or pointing in a direction, they are a quantum smear; they just have an energy that depends on the direction and magnitude of the magnetic field that affects them.

Nobel laureate Wolfgang Pauli was the first scientist to propose the basic concept of spin in the 1920s. A collective of physicists based at Leiden University proposed in 1925 a 'physical interpretation' of particles spinning on their own axes, but it was Pauli who crunched the numbers so as to articulate the mathematical theory to support the hypothesis of particle spin in 1927. There are many unique characteristics of atomic spin that further distinguish it from its macro-equivalent produced by the mechanics of gravity. For example, while spin theory holds (and observes) that particles can change the 'direction' of their spin (the behaviour of axial angular momentum), the rate at which any spin occurs is always relative; a particle's spin 'speed' is dependent on the external magnetic field affecting it.

Particle spin is a fundamental value in quantum mechanics because spin theory, as substantiated by work across both experimental and theoretical physics, suggests that the spin of particles cannot be explained by appealing to the hypothetical existence of particles even smaller than fermions and bosons. That is to say, the behaviour of particles cannot be even hypothetically explained through models in which the smallest known particles are themselves constitutive of even smaller particles, as the axial angular momentum of particles cannot be accounted for in such models. Therefore, atomic spin theory has established that, within the bounds of current knowledge, subatomic particles are truly the building blocks of physics as they possess 'no inner structure'. This means that the spin value of elementary particles is, according to current models in quantum theory, a truly intrinsic physical property of the subatomic universe.

To explore the functional possibilities of the strange quantum mechanics of spin is to play the part of pioneer in the little-known fundament of the physical world. Of course, the vanguard pioneers of this enterprise are those physicists, like Pauli, who first identified the possibility of this fundament existing at all. Today's particle and quantum physicists are undertaking experiments that are unprecedentedly massive and underwritten by billions of dollars. When we think of particle physics today we don't simply think of blackboards in musty lecture halls coated in equations, but of the construction and firing of the Large Hadron Collider and the subsequent revelation of the hypothesized Higgs boson.[1] At every major paradigm shift, artists and creative thinkers have welled up, perhaps first as a trickle, but soon as a sizeable guard – enabling contestable science or unintelligible proofs to find some critical footing in society. The rise of the quantum world is no exception.

The creativity of the spin

The discovery of the quantum world allows for an immersion into a field of physics that is directed not by classical empiricism, but by the ruptures of probability. In the quantum world, creativity finds a strange familiarity. Perhaps this resonance is due to the common ground of uncertainty. In quantum physics, the uncertainty principle is not simply a by-product of clumsy observation (of intrusive laboratory experiments) but rather refers to a physical, observable, verifiable un/property of our physical world. This revelation of the physics of uncertainty restores, it seems, something of the unknowable into the everyday and collapses some of the monopoly of knowledge into the realm of imagination and creativity. It is the concept of controlling a natural intrinsic quality in the world that makes the spin such an area of intrigue. The spin that is not a spin is unlike science in the spotlight of Modernism, and its colonialist sense of progress and reason. Quantum theory and its correlate fields of physics in effect defy both the mechanics of common sense and the idealism of Enlightenment thought.

The affected atomic spin, swerve, swirl, eddy, vortices, rotation, turn or gyration, is the manifestation of the creative immeasurable. Qubit is the term for a single quantum bit, a quantum analogue of the classical 'bit' but which is innately possessed of two potential states. The curious nature of sub-atomic spin is that qubits become entangled with each other and form 'quantum-entangled states'. The result of this process is those particles, which are entangled with each other, cannot be understood as either 'up' or 'down' (corresponding to integer or half-integer spin value) but as existing in a number of possible states. Entanglement happens when you have the spins in superposition, so that the measurement is probabilistic – and the outcome of measuring one spin determines the outcome of subsequently (or simultaneously) measuring the other. So the superposition up-down + down-up is an entangled state, because you don't know in which state either spin is, but after measuring one you know with 100% certainty that the other one will have opposite spin (even if they're infinitely far apart). Likewise up-up + down-down is also an entangled state; again you don't previously know the result of measuring either spin, but after you do you're certain that the other will have the same spin.

The qubit, as stated, is a unit of quantum information analogous to the classical 'bit'. Each qubit is in itself a two-level system encompassing a set of alternate states. Thus, a qubit is a two-level quantum mechanical system. A single photon, for example, constitutes such a system. The difference between a classical (non-quantum) two-level system and a qubit is that, in a classical two-level system, its object (bit) would either necessarily be in one state or another: up or down, left or right, etcetera. In quantum two-level systems, the object (qubit) exists in a superposition that straddles all possible states in the system (that is, in a sense, both up and down, left and right, etcetera). This unique property of the qubit to exist in more than two quantum states simultaneously is the functional basis that makes quantum computing possible. The reason for this is that while a bit is limited to a single packet of information ('0' or '1' to use the binary code example), a single qubit may contain the information of two classical bits insofar as its superposition means you only obtain one of the eigenstates as a result.[2] The opening up of quantum two-level systems as an information technology used in computing is achieved through a process called superdense coding.

Of course, the concept of spin that quantum physics borrows from the classical world is only used as an analogy, necessitated by the counter-intuitive and abstract universe of the sub-atomic and quantum. Michael Serres fleshes out this conceptualization of quantum spin by reiterating and expanding upon a passage from Plato, who wrote about the swerve of a spinning top as analogous to the deviation of the human soul from nature. Serres writes, '[the spinning top] is in movement, this is certain, yet it is stable [...] its axis may lean, take on an inclination, without putting the movement of the whole in too much danger' (2000: 29). So while the axial momentum of the spinning top dissipates in the classical world, the axial momentum of the quantum spinning top (the particle) has a fixed rate, even if its direction may change. It is the orbital deviation of the quantum spinning top that expresses, oscillates, leans, sways, rotates and mutates; it is this swerving motion that expresses 'all the movements known and thinkable' (2000: 29).

The metaphorical fabric of our universe is revealed through interconnected individual fibres and strands, which were originally spun into threads and in turn are woven together to become the fabric. Spin is then, on a material level, intrinsic to the metaphor of fabric as being the space of the void, dark matter and the universe.[3] The cosmic fabric is therefore part of our own personal spatial layering – it envelops us in the world.

The issue is that the physical spin referred to in a classical world does not fit into the quantum world where phenomena happen at the very smallest scale. At the heart of quantum mechanics is quantum paradoxical spin; it is the most fundamental (known) quality of the atomic world. Current orthodoxy in quantum physics holds that the behaviour of subatomic particles cannot be accounted for by appealing to yet-undiscovered particles. In other words, the base quantum mechanics of spin is an inferred quality of the very fundament of physics as we know it, because in current models, this quality is understood to be innate. Therefore, Serres' thinking on quantum spin and the deviation of the human soul is an analogy between the most basic precession of matter and the (unnatural) crux of the human condition.

Given that the spin value of subatomic particles is not a character of particles but, rather, the very quality that determines and defines particles to begin with, the physical world as we know it can be most accurately understood (using the most developed scientific theory available) as the flux and angular momenta of quantum matter. This strange world of resonances produced by the oscillation of myriad particles is one that was conceived of in the infancy of quantum theory. Take this passage from Feynman in 1982:

> It's been noted time and time again that the phenomena of field theory (if the world is made in a discrete lattice) are well imitated by many phenomena in solid state theory (which is simply the analysis of a latticework of crystal atoms, and in the case of the kind of solid state I mean each atom is just a point which has numbers associated with it, with quantum-mechanical rules). For example, the spin waves in a spin lattice imitating

> Bose-particles in the field theory. I therefore believe it's true that with a suitable class of quantum machines you could imitate any quantum system, including the physical world. But I don't know whether the general theory of this intersimulation of quantum systems has ever been worked out, and so I present that as another interesting problem: to work out the classes of different kinds of quantum mechanical systems which are really intersimulatable – which are equivalent – as has been done in the case of classical computers.
>
> (Feynman 1982: 475)

The reference is important and raises the issue of a paradox between the real and simulated. Here, Feynman not only models the world as a series of quantum spin-waves but goes on to hypothesize that the data could, given powerful enough quantum computers ('quantum machines'), be imitated or replicated via computational processes. In this quote, Feynman is hearkening in the quantum age whereby the atom is no longer modelled as a classical object but as numbers relevant to quantum mechanical rules. Feynman's formative seminal paper took the intuitive leap that saw the development of quantum computation and, in turn, my current interest in research being carried out at the Centre for Quantum Computation and Communication Technology (CQC2T) at the University of New South Wales (UNSW). It was through my relationship with the CQC2T at UNSW, notably the research and experiments being undertaken by Andrea Morello, Juha Muhonen and Juan Pablo Dehollain, that I gained access to first-hand data on the science of controlling the spin of electrons (2010).

Since 2010 the CQC2T team have been working with a perfect crystal of silicon-28 (material can be sought from off-cuts of the International Avogadro Project).[4] This unique silicon (Si) is the element with atomic number of 14 which means an atom has 14 protons and 14 electrons. The electrons are structured into three energy levels, which have two, eight and four electrons (from inner to outer). The four electrons in the outer levels are the ones that bind to other atoms, and this is why Si is tetravalent, because each atom in the crystal is bound to four other atoms via their outer level electrons.

Phosphorus is the element with atomic number of 15, with electron energy levels two, eight and five. When a phosphorus atom binds in a Si crystal, it uses four of its outer electrons to bind to four adjacent Si atoms, which leaves the fifth 'loosely bound' electron, which can be manipulated by external fields to be used as a qubit. In the experiment, a pure silicon crystal is held at sub-zero temperatures and in a vacuum state that allows for silicon's electrons (four) to bind, like interlocking 'teeth', with a phosphorous atom, as a silicon solid state. The extra phosphorous electron is held in a magnetic field to maintain its spin in a single, controlled direction where it can be manipulated through electromagnetic pulses (Fig. 21).

To put the experiment described above in the context of our running analogy: imagine negative charged electrons as spinning tops who are attracted by the nucleus positive charge (whose axes are fixed) such that while the first spinning top is still free to spin, that gyration is now able to be manipulated through the transmission of electromagnetic current (microwave signal). Trapping a single phosphorous in this atomic/solid-state structure allows for the manipulation of quantum entanglement – which innately links quantum particles – opening up remarkable possibilities for quantum coding (and hence quantum computing). It is a strange phenomenon to control the single atom, to make it perform to our wishes, to extract it from its normal environment, entrap it – to control what might not have been considered significant now, post the nuclear age.

Figure 22: The Fall, 2018 Digital print, made in processing, from raw data based on the microwave signal effecting one Phosphorus electron; 800 microwave separate 5 milliseconds signals.[5] Courtesy of artists Paul Thomas and Kevin Raxworthy.

Quantum consciousness

In the first iteration of the *Quantum Consciousness* artwork, I sought to internalize a contextual and conceptual framework to research for an artwork based on data from the CQC2T team that would enable me to process and transmit ideas from a personal perspective. The concept developed created an actualization of the information into an audio-visual artwork of quantum superposition states and the probability of their positions (Morello et al. 2010). Classical bits are either zeros or ones, but qubits (quantum bits), as discussed, can be in both states at the same time. Every individual qubit (particle) can, like a classical bit, either be 0 or 1, or any quantum superposition of those two qubit states, but when entangled with other particles assumes a new quantum value that is neither one state nor other, but rather a probabilistic average of all – possible states. The concept of entanglement is a physical phenomenon that is produced by groupings of particles (pairs or greater), which create a collective quantum state whereby each individual quantum unit cannot any longer be described independently. In other words, it is when the information contained in each electron is co-constituted and flows between each other. I sought to give form to this quantum state and allow the viewer/participator to experience a sense of immersion in the material properties of quantum phenomena.

The quantum phenomena of entanglement have parallels with the work of creative artists produced since the mid-nineteenth century; artists and thinkers have continually questioned the empirical certainty of the object-world. This line of questioning is of course commonplace in religious metaphysics, which is why I suggest the nineteenth century as an important rupture, with the onset of the Enlightenment and the cultivation of atheistic humanism in Western Europe. In this time, we see the first ruptures in the perspectival-dominant gaze of Western art and a rethinking of what it might look like to observe and to attempt to comprehend the world from multiple positions, not simply from a single and fixed position. In other words, I am here analogizing multi-perspectivalism or the twilight of perspectivalism with the emergence of quantum theory. This is the heart of it: developments in the visual arts (intelligible as the making of marks towards the understanding of the world) can be aligned with lessons from quantum physics that complicate the existence of empirical objects and perspectival certainty by reducing their latent materiality to theoretical probability, as if to play into Cartesian fears about the insufficiency of empirical study.

Here we are comparing creative imaginings, representations and imaginings of objects in space with the natural, physical state – not of the classical world – but of the quantum world. The part that the artist plays as an observer is where this particular problem is highlighted. The artist, who effectively seeks to test the world through his or her models, is attempting to give form to intuitive questions that are natively difficult to express. This is to say that artists have, in some way or other, for hundreds of years been attempting to bring to light what contemporary quantum physics (since the early twentieth century) is now able to show us. In other words, we are exploring the art of uncertainty.

One of the initial premises of this line of thought stems from epistemology and the paradox of how new knowledge is formed. Here, we are delineating a trend in thought that has intuitively rejected perspectivalism (and its correlate 'knowns'). It seems improbable that we could survive with not accepting or acting on what we know. The very act of attempting to define some concrete thing is to create a ground from which to base an opinion. This opinion or model has to be seen as emerging from a sum of probabilities. The sum of probabilities, where we draw our comprehension of the world – through experience, discernment, reflection and critique – can't be seen as fact; we can never see the myriad possible worlds but only the one we choose to see. The probability concerning the factuality of a given model of the world is always in conflict with the probability of another (competing) model, and it is for this reason that in the brain, models of the world struggle against each other (from subconscious to conscious) for territorial dominance, for certainty. The ground of thought is therefore uncertain, like the shifting sands of the desert: an inherently unstable platform of shifting meanings and contexts. When the competition is effectively concluded and a model of the world is declared the victor, this has real effect, collapsing all other possible worlds into a single monopoly. If this process, which we call the experience of 'certainty', is allowed to occur, and cognitive dissonance therefore checked, we have arrived at a classical model of the world. But what if we were to look to a model of the world that is in itself innately unstable: a model that explicitly notes its own insecurity and even impossibility as a 'true' and certain model of the world? This is exactly what we mean to explore here – the reality of quantum phenomena to justify initial uncertainty as the foundational model of the world.

In attempting to control and understand the spin of a particle, there needs to be evidence of what can't, in quantum terms, be evidenced. In other words, in quantum probability theory there exists the latent inability of measurement to predict, before it is observed, whether the direction of particle spin will flow north (up) or spin south (down). It is an empirical fact, in quantum mechanics, that uncertainty is a fundamental rule of the physical world; certain physical processes can only ever be summarized, never predicted, given the implicit uncertainty of the quantum world. It is important to stress and reiterate that this problem is not a shortcoming of observation, that is, quantum uncertainty is not a stopgap necessitated by imperfect analytical and observational methods – quantum uncertainty is a physical property of the material universe.

Experimental research can indicate when an electron is in spin up or spin down, but any observational measuring process can un-entangle quantum entanglement. If particles become entangled with each other on account of quantum mechanics, those particles now physically exist in an uncertain state, known as a superposition. Therefore, by definition, superposition is a counter-intuitive quantum state that is classically measureable only because it describes a field of probable classical states. In my work, I have tried to imagine being inside the probability field of a qubit, mapping/visualizing its domain with a longitudinal and latitudinal wire frame to create a probability 'globe' (Fig. 26). Positioned within this probability 'globe', what do we hear? What formal properties can we perceive, if we were small enough, in the materiality of quantum uncertainty?

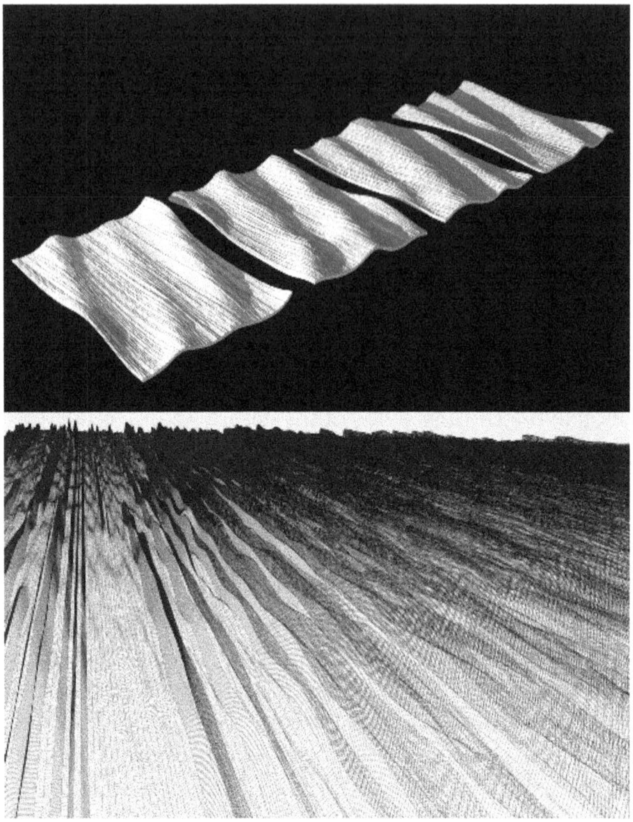

Figure 23: Kevin Raxworthy, 3D model (above) and digital image (below) of selected probability data from the microwave signals affecting a spin of a phosphorous electron. Courtesy of artist Kevin Raxworthy, 2018.

My current working interest is in taking and manipulating, or visualizing, the fabric of quantum space-time to form new visual interpretations based on the resonant parallels between quantum phenomenon and creative intuition. By taking license with, and pulling concepts up from, the infinitesimal scale of the quantum and inserting it into the familiar domain of the classical (macro) world, I intend to show and explore how this creates confrontation between knowing and not-knowing.

In the aforementioned work, *Quantum Consciousness*, I worked in collaboration with QCC. Data was provided by QCC concerning the microwave signal effecting the raw material of atoms to extract out the probability data to ascertain an approximation of a position. This raw data was explored and visualized in a number of different mediums to develop the conceptual basis for the immersive artworks (Figs. 23 and 24). The data was the feed for my imaging and sonification of the superposition of a quantum particle. This microwave signal

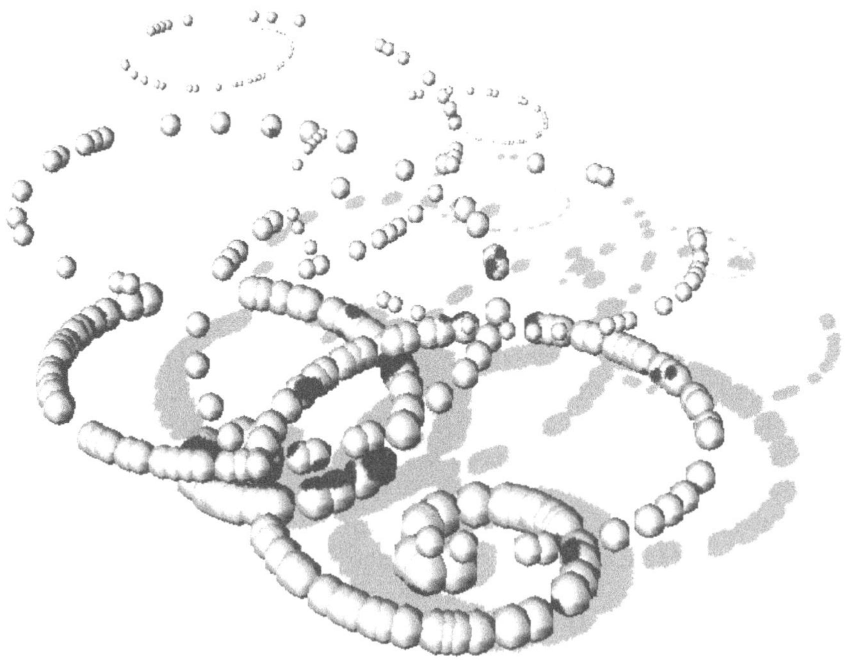

Figure 24: Illustration demonstrating the probability of the electron's positions from being affected by the microwave signal. © Paul Thomas.

was, in the initial installation, applied to a reading of Richard Feynman's 1982 paper on the birth of the quantum computer, which I have cited above. This reading, in its transformed state, stimulated the probability movements of an electron-as-superposition. Applying real quantum research methodology, this work maps the layering and averaging of data to establishing and narrowing the probability field of a particle position, and expresses this possibility through the voice of Feynman. Feynman's voice is mapped to a set of speakers to imitate the immersive field of probability, and is heard (in broken fashion) as coming from one place or another depending on the data-stream from CQC2T via our probability equation. In effect, my work is an audio-representation of what in quantum physics is known as a Bloch sphere.

In *Quantum Consciousness*, the electron from which the observational data is garnered is controlled by a magnetic field that needs to oscillate at a very specific frequency. Whilst attempting to understand the data referenced by Morello, a series of computerized drawing stimulations were created of the microwave signal that effects the electron to spin to its superposition. In Fig. 24, the viewer can see patterns that influence the spin, where the 10,000 points in each line of data references that vertical fall in the void to show where and whether ticks might occur. The microwave signal emitted stimulates the electron to spin, therefore causing the most miniscule of movements for the electron. This is provided by the aforementioned microwave source. As long as the frequency of the microwave is stable at the right frequency, the electron will move monotonically around the immersive Bloch sphere: my sonic and geometrical representation of the qubit. The better defined the qubit's position is on the Bloch sphere, the more coherent it is; if it decohered, its position would get completely lost (e.g. it can't track it anymore) and the experiment would fail.[6] If the microwave source is only slightly off-resonant, it will vary the speed at which the spin rotates and also around a distinctly different axis, compared to the 'ideal' axis-of-movement when the signal is resonant.

Talking to an electron was the second iteration developed in relation to the *Quantum Consciousness* project. *Talking to an electron* used my voice in reading selected excerpts of Feynman's 1982 paper to affect a microwave signal that stimulates and excites an electron to spin. Reading Feynman's paper to the electron enables a loop of thought from the instigator effecting what has been instigated at its fundamental level. Feynman states to the electron that 'because nature isn't classical, dammit, and if you want to make a simulation of nature, you'd better make it quantum mechanical' (1982: 486). The sonic microwave signal is used in quantum computer research to excite the phosphorous electron to spin to its superposition. The microwave data blips are registered on an oscilloscope and this data is used to drive the sonic and visual expression of the quantum computer processing the thoughts of Feynman. These blips can thereby be understood as representing the 'geographical' position of an electron (probabilistically) within the Bloch sphere: triggered in the value of the spin and the frequency of the blip.

The need for a conceptual framework for the data, based on affecting the spin of the electron, enables the artwork to develop a poetic sense of having a human effect in the quantum world (Fig. 25). By feeding the signal from my voice recording into the frequency modulation input of the source, the frequency of the microwave is altered by small but perceptible amounts. Therefore, *Talking to an electron* expresses user-effected quantum permutations revealed as dynamic changes in the precession of the electron within the Bloch sphere.

As for the sampling rate, Dehollain[7] from CQC2T took 1000 samples of the qubit precession in total. This had to be done within the lifetime of the qubit, so in the experiment these 1000 samples were taken across the duration of 60 microseconds. In order to accomplish this, Dehollain had to resample the sound signal so that the minutes-long voice recording of participants could be compacted to an equivalent duration (this is the same with the video-representation of the experiment). The measured current from the special silicon transistor enabled the observation of the spin of the electron. When the atom spin is valued as 'down', there is no current – and when it is 'up', there is a blip. These blips are important because they serve as the trigger points for the audio-visuals. The data from the blips is registered on an oscilloscope and sound card to be played back through audio speakers. These blips can thereby be understood as representing the 'geographical' position of an electron (probabilistically) within the Bloch sphere, triggered in the value of the spin

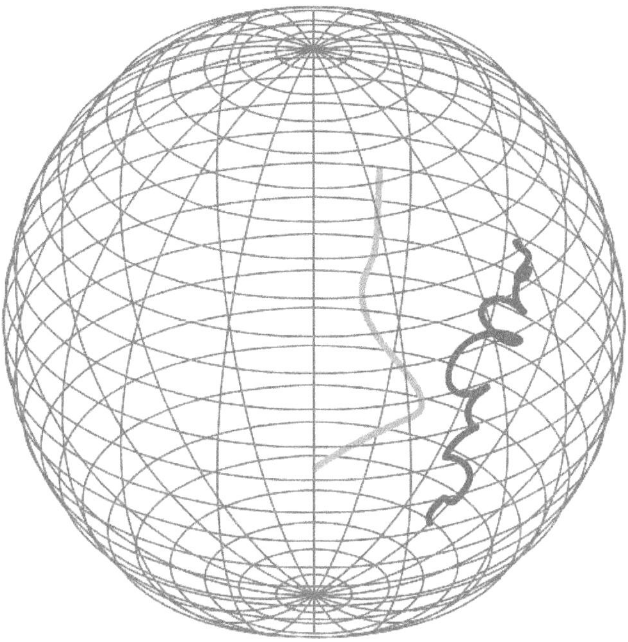

Figure 25: Diagram showing dark grey and light grey lines. The light grey line shows the wiggles in the experiment that are consequences of the voice changing the path of the rotation of the spin, but the spin is always coherent in the experiment. Graphic by Paul Thomas, 2018.

and the frequency of the click. In *Quantum Consciousness*, one can listen to the curve of the particle as it spirals through a longitudinal and latitudinal diagram within the quantum 'globe'.

Quantum Consciousness (Fig. 26) is a visual/sonic installation that immerses the viewer in quantum phenomenon. The artwork is an audio-visualization of atomic data that indexes the quantum 'movement' of an electron. This project delivers an aesthetic and immersive experience that places the viewer inside the virtual superposition of a sub-atomic particle (i.e. its field of probability). This audio-visual experience establishes a metaphor and indexical map of human thought. The quantum spin of the electron, mapped here, is serving as the conduit for the form of the sensory stimuli of the work. The broader significance of *Quantum Consciousness* concerns the possibility of non-human, computational intelligence commingling with human consciousness. This link stems from the fact that the 'material' of the electron and its quantum behaviour is the basis of the possibility of quantum computation.

The data that is *Quantum Consciousness* can be thought of as entangled with the artwork and the totality of the elements that makes the work. Each part is entangled with the next, leading down to the very words recited by Feynman to the electron. In the

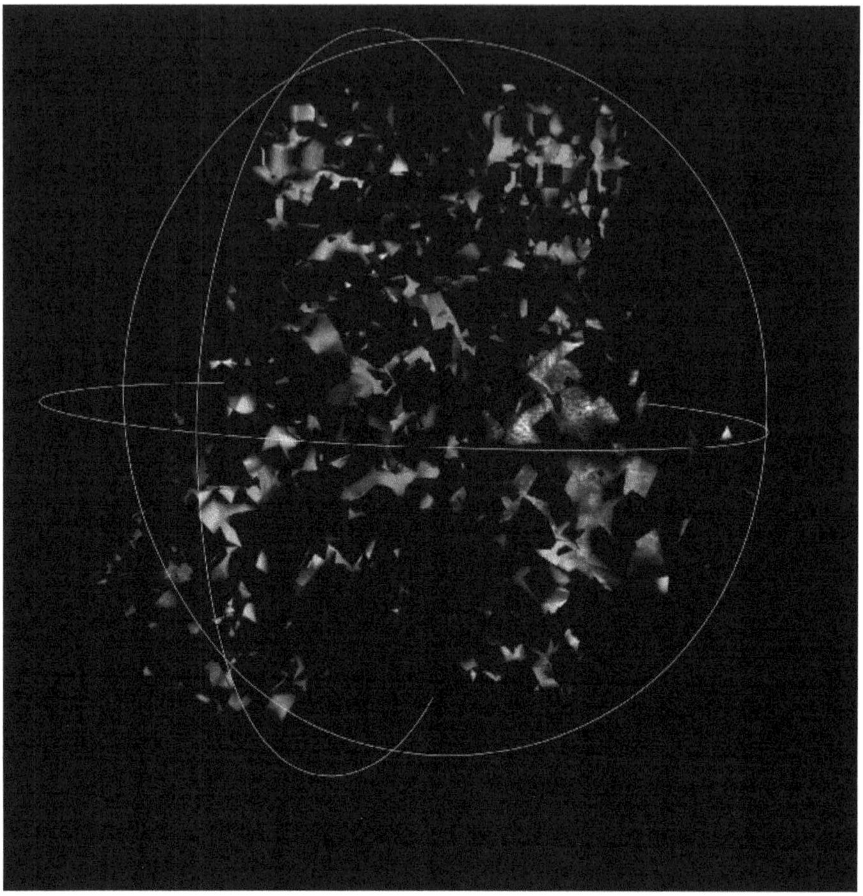

Figure 26: Paul Thomas, *Quantum Consciousness*, 2016. Screen capture showing the orbital positions taken from the probability data. © Paul Thomas.

context of *Quantum Consciousness,* the spin value of the interrogated electron (a quantum mechanical property) is imaged in a way that does not only parallel the classical world but, through its murmuration and the expression of the ebbing and flowing of physical probability, represents, in a fashion, the sum total of what could be – of the plethora of possible emergences that subsequently constitutes the 'classical' materiality of the world. Therefore, the concept of superposition can be understood as having resonant links with the developments of conceptual art in the 1970s, given that it was the imperceptible positions or context for ideas (invisibly structured) that served to generate, through a signifier, that which was indexical to the artwork. Furthermore, the link between the quantum computer, consciousness and artistic expression is emerging (and which I aimed to image in *Quantum Consciousness*) through the seeming co-emergence of thought and quantum conditions.

Quantum Consciousness, both the artwork and the concept, is an experiment in the imaging and materializing of impossible (quantum) states of matter and the co-emergence of human consciousness.

Consciousness and the swerve

Let me expand upon this function (latent in conceptual art processes) of context as a triggering device, wherein the viewer observes the world that emerges and is created through a given context, and which generates the specific signals being sent from the eyes and, subsequently, firing neurons in the brain. These signals can be seen as analogous to microwaves affecting the spin of the electron in the brain of my work *Quantum Consciousness*, and furthermore can be understood as a classical analogy to the 'consciousness' of a quantum computer. In this respect, artists play a role in taking up the potential of what Lucretius referred to as the clinamen – a concept he borrowed from Heraclitean atomism and which refers to the unpredictable swerve of atoms in their movement through space. The interactions and 'crashes' caused by atomic swerve, by the clinamen, is therefore the trigger-point or material rupture through which cosmic events manifest. According to Lucretius, were it not for the unpredictable swerve of atoms, not only would there be an absence of free will (given the stranglehold of determinism) but there would be no objective universe (as we know it), given that none of the atomic interactions necessary for material complexity would ever take place.

The clinamen is the bridge between the invisible processes and mechanics of the atomic world and the evident and perceptible materiality of the classical world of objects and form. The clinamen is, as such, that self-same blip of data, that kink in the fabric that brings things into existence to be experienced in an objective fashion. Equivalently, the conceptual work of art creates a way of engaging with the material phenomena of energy and atomic vibrations from the conceptual plane of intuition to the material plane of objective experience. It is the initial uncertainty of what is revealed to the brain (by the procedures of rational human consciousness) that suggests the very operations of cognitive intuition are when they first appear, in that moment predicted and hearkened in the early stages of the development of the quantum computer. In other words, the currently emerging development and realization of the first quantum computer is an anthropocentric reflection of humanity's comprehension of themselves as a machine.

What is comprehended by the material structures that are non-human elements that no less constitute the condition we call 'human'? How can what is being expressed by the non-human (the natural)/human material construction be comprehended by the human subject and then expressed to others? Empiricism is in many ways a law of averages (borne of repeated statistical observations). By the same token of empirical observation and averaging are quantum superpositions narrowed in upon and through the same process perspectivalist art realized through comparative analysis with the physical world and its laws. The means through which a superposition is gleaned is not a process that is unique to quantum theory – it is a fundamental rule of statistics.

Atomic spin, with its internal rhythms and vibrations, is in many ways indexical to the material characteristics unique to various artistic and creative mediums. Materials, irrespective of being natural or artificial – such as graphite, oil, metals, wood, cloth, plastics, etcetera – at their atomic core, reveal intrinsic agencies and properties in such a way that any distinctions that are self-evident in the world of classical objects no longer apply. The ability to 'envision', that is to attempt to model the world through materials, is not simply to see their anthropocentric possibilities – but rather to appreciate and listen to the strange familiarity that those chemicals, which constitute our human-biological make-up, share with their equivalents in the external world. The mediums of creative expression we have cultivated, and so inherited (as a species), over thousands of years trace gestures and intuitions that communicate (and are communicated by) a transversal arrangement where all parties are interdependent and therefore vital. The energy latent inside the unique atomic arrangement of every material is equal to the energy that drives the agency of that material. The quantum world is built on uncertainty, magnetism, gravity and forces of attraction that are perceptible through rhythm vibrations and patterns. So, these forces are inherent to each independent atom, whether it is part of the human or the non-human material of art.

The concept of the superposition of particle spin is significant in the line of thought articulated in the preceding paragraphs because it allows for, and reveals, that the probabilities of things exist in an undefinable state. There is the potential, latent in the quantum world, for a series of 'becomings' that parallel the inherent indeterminacy of superposition borne of quantum entanglement. Here we can see the links that circumvent the Anthropocene and flow from and between the cosmos (material world) and the uncertainty of the quantum world.

Art has a power to frame the imperceptible forces of chaos and uncertainty and to amplify aspects that make the infinitely small and imperceptible phenomena of the universe tangible. It can take the instrumental 'machinic' interpretations of science, along with the poetics of the conceptual, to reinterpret the very agents that incorporate the forces, vibrations, rhythms and flows that constitute our universe. These forces are the mediums (to use an artistic analogy), and serve as the constituents and the material origin-points for imaginative 'becomings' in the arts. Thus, a given medium and physical property is the agent of a 'becoming', meaning that those material chemical compounds that are the basis of the human are also elements in and of the art material.

The quantum computer-human relation is made strangely real in this quote by Deleuze, first published in 1988 in his book *Foucault*:

> Is it not commonplace nowadays to say that the forces of man have already entered into a relation with the forces of information technology and their third-generation machines which together create something other than man, indivisible 'man-machine' systems? Is this a union with silicon instead of carbon?
>
> (Deleuze 1999: 74)

The experiments using silicon atoms to metaphorically cuckoo-nest[8] a phosphorus[9] electron that becomes the substrate for the fabrication of the thinking man-machine systems. The intuitive understanding on materiality reflected in Deleuze's writing enables us to see the vulnerability of humans 'being supplemented or altered by means of a material prostheses' (Marks 2010: 159).

The predictions for the role of silicon as a material that overshadows that of carbon, as part of an evolutionary process, opens up the discussion to the concept of temporality in the human/non-human genetic material agency. Current experiments with carbon nanotube demonstrate that this material appears to be fighting back. The idea of investing in one precise materiality to make the analogous relationship between human and machine thinking is more complex than simply mirroring existing solutions. We are prone to invest in materials that become core in established, impregnable, corporatized institutions. These institutions are problematic to reverse, even when the environmental inappropriateness of the material is revealed.

The prosthesis in the context of the quantum computer is questioned by the very nature of what is being established about what constitutes agency or life in non-human objects. Is it the energy of the spin, the unpredictability of the swerve or the materials that conspire together, forging an indivisible feeling to make and create material consciousness, that needs to be cognizant to have its own life?

Notes

1 The Higgs field can be found throughout the universe and is made up of particles called Higgs Bosons. Imagine describing the Higgs field as water in which neutrons, protons, quarks, photons and electrons all swim through. Each particle moves through the Higgs field in different ways. Some particles are streamlined and move through without contacting Higgs Bosons, whilst others create more resistance and therefore gain more mass. It is the very existence of the field and its relationship to Lucretius's void that causes all matter and nature to evolve.

2 In a classical world, we can say that every object appears to have a definite position, a definite momentum and a definite time of occurrence. In contrast to this, the uncertainty principle says that it is impossible to measure the momentum of a particle if its position has been measured. When a particle can definitely be measured in some respect, it is said to possess an eigenstate.

3 The definition of spin comes from spinning in regards to fibre, and means to draw out and twist fibre into yarn. The drawing out and twisting is particularly useful to bear in mind as a creation of the threads for the fabric of the universe.

4 The International Avogadro Project has been working to describe this number with sufficient resolution. Instead of a chunk of carbon-12 atoms, they use a uniform crystal of silicon-28, carefully machined into spheres with a mass of 1kg based on the current definition ('Redefining the Kilogram: Silicon' 2014).

5 Whilst attempting to understand the data referenced by Professor Andrea Morello, a series of computerized drawing simulations were created of the microwave signal that effects the electron to spin to a superposition. In this drawing the viewer can see patterns that influence the spin where the 10,000 points in each line of data references that vertical fall in the void to show whether spikes might occur and where. The microwave signal emitted stimulates the electron to spin, therefore causing the most miniscule of movements for the electron.

6 Coherence and decoherence are important for control in the development of quantum computers that need the states of the electrons to always remain coherent. Decoherence is the outcome of an aspect of quantum phenomena developing irregular oscillations. In *Talking to an Electron*, the diagrammatic trail evidenced in the Rabi oscillator (a method of measuring the spin of an electron named after Isidor Isaac Rabi) reveals the problem and subsequent collapse of the electron losing its quantum state.

7 Notes sent to author from Dehollain: 'I fed the sound wave from your voice recording into the frequency modulation input of the microwave source that controls the qubit. I'll try to explain a bit better what this means: The qubit is controlled by a magnetic field which needs to oscillate at a very specific frequency, is supplied by the microwave source. As long as the frequency of the source is stable at the qubit frequency, the spin will precess monotonically around the Bloch sphere, from down to up to down and so forth. If the microwave source frequency changes by a large amount (i.e. the microwave source is off-resonance with the qubit), the magnetic field no longer has any effect on the qubit and the spin stops moving in the Bloch sphere. Now if the microwave source is only slightly off-resonant, it will make the qubit precess at a different speed, and around a different axis than the "ideal" perfectly on-resonance case. By feeding the signal from your voice recording into the frequency modulation input of the source, I am changing the microwave source frequency by small amounts, which are guided by the signal in your voice recording, therefore the qubit displays dynamic changes in its precession accordingly'.

8 The cuckoo bird lays its eggs in the nest of other species of birds.

9 The possibility that quantum processing with nuclear spins might be operative in the brain is explored. Phosphorus is identified as the unique biological element with a nuclear spin that can serve as a qubit for such putative quantum processing – a neural qubit – while the phosphate ion is the only possible qubit-transporter (Fisher 2015).

Chapter 4

The Graphene Moment

The genealogy of graphite

This chapter locates the historical moment of the discovery of graphene (2004) as a seminal catalyst in the new materiality of the twenty-first century and its object-oriented ontology.[1] The discovery – or more appropriately, the incidental invention, of graphene serves as a springboard for a revisionary critique, challenge and exploration of the privileging function that direct human (sensory) experience holds over non-anthropic objects. In this moment, the ubiquity of graphite (an essential formation of carbon) was usurped and hidden material agencies were brought to light. This reconfiguration potential of material agency is now, in an age of post-atomic certainty, the defining feature of new materiality.

New material possibilities (of discovery within discovery) mark an evolution towards the possibility of quantum matter.[2] It is quantum matter that sits at the horizon of current developments in new materiality. In this future, the quantum nature of matter has altered all certainty into a sense of latent unpredictability. The graphene 'moment', in which we now live and think through, all began courtesy of centuries-long examination of and through the humble graphite pencil. The serendipity of new (revisionary) materiality is that, in this case, the graphite pencil served as a necessary historical vehicle towards the discovery of a new agency within itself (carbon) – that is, graphene.

The tool of the pencil and its material medium (graphite) serves as the basis of this chapter. The graphite/graphene agency functions as a provocation to reconsider how we understand the material world and our syncretic relation to it. The construction of questions related to an artistic interpretation of the world facilitated by graphite is fundamentally significant now (in the graphene moment) to destabilize and offset the quantitative certainty approach of atomic science (Fig. 27). The properties and potentiality of material agents, once outside 'the human', are borne of natural mechanics (physics and chemistry) where the free agency of matter rules. Human appropriation of materials dulls and dims them and disguises the latent properties of materialness as they exist in nature, ignoring the power of intuition as a force for enabling material agency to come into existence.

A change in intentionality for matter, when viewed in relationship to what we now understand from evidence presented by art and physics, becomes a non-materiality predicated upon a quantum-oriented ontology.[3] The agential[4] properties of any material can change its (applied) meaning, form and value when what is discovered, uncovered or revealed of the new-found properties is linked to a social and scientific developmental process. Scientific (analytical) technologies allow us to physically unlock new agencies in materials already appealed to, for a previously uncovered agency. The basic chemistry and materiality we unlock (in the search for new agencies) is the same chemistry within our own bodies – meaning we 'unlock' new intra-actions[5] fundamental in the chemistry of our own biology.

We often see our existential (human) relationship with material objects shift and recoil from a strange dissatisfaction with appearance or shifting social standards in relation to the world of objects. Shifting of the standard of world views manifests through the ideologies of form, style and aesthetics. Therefore, our views of the world are intrinsically linked to the desire for material consumption. Anthropocentric exploration and geographical mapping of the planet by mostly Western powers in and around the seventeenth century meant that for the first time the finite nature of material resources was understood. In other words,

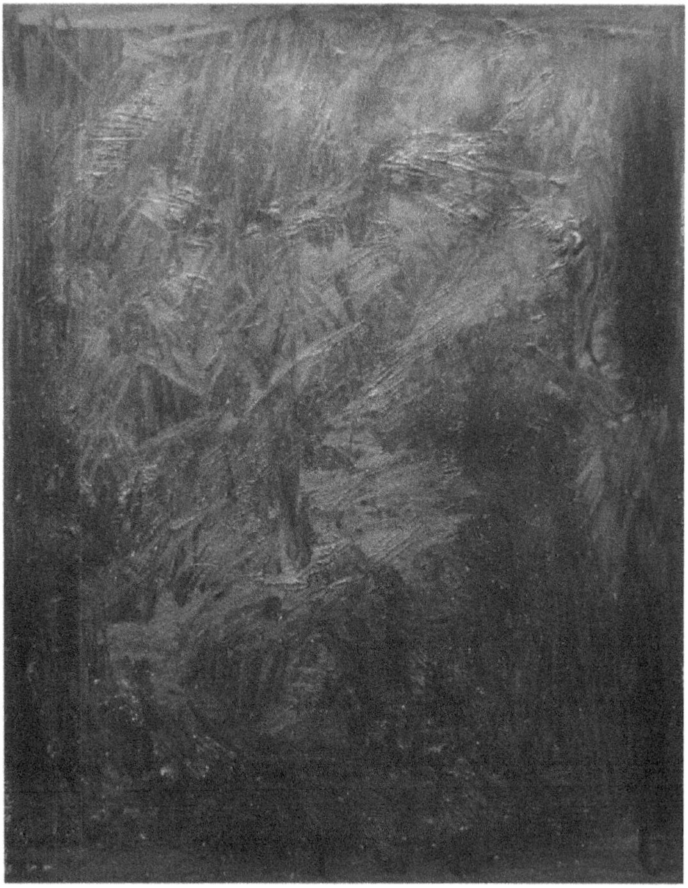

Figure 27: Paul Thomas, *Quantum Spin*. Graphite on paper drawing of the sonification of a microwave affecting an electron's spin, 2014. © Paul Thomas.

colonial powers realized that they had to take control of and aggregate a material wealth to ensure the possibility of future global (colonial) projects and nation-building.

The seventeenth-century realization that the (physical) world is a finite composition of specific materials meant that certain uncommon materials became inherently valuable on account of their central importance to various industrial and scientific enterprises (ores, metals, etcetera) (Fig. 28). Not long after the seventeenth-century colonization of the physical world of objects, came a period of colonization of the atomic world.[6] This was achieved through a molecular discovery of material agency and became another major means of reifying empire and power. The colonization of the atomic world has led to the construction of new materials through nanofabrication processes and opened up a whole new world to explore and map – even beyond the atomic. These developments of new materialism echo those much earlier chemistry experiments where scientists combined materials together in search of the discovery of new material properties. Because this resultant set of unique

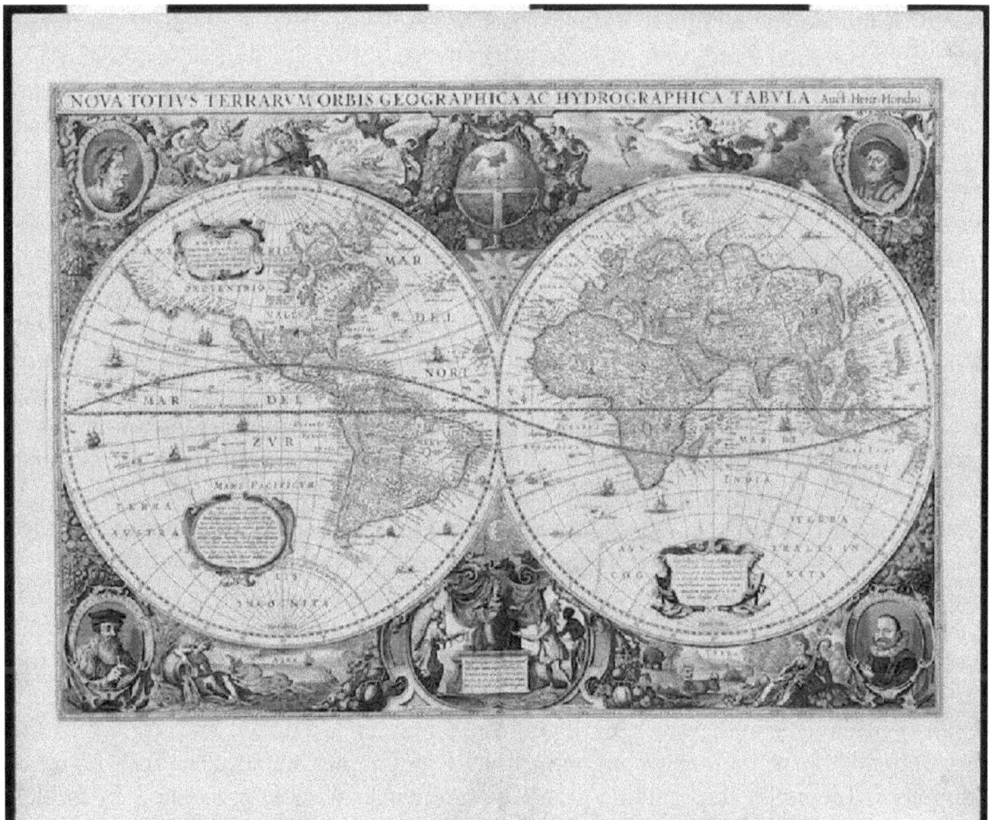

Figure 28: Hendrik Hondius, World Map, 1641. Photograph in public domain.

materials and material agencies discovered through the experimental mapping of the atomic frontier was in itself facilitated by the application of older and more obvious material agencies, there is, in the age of new materialism, a sense of materials discovering aspects of themselves. In this context, we are situating the relationship between graphite and graphene as a necessary synthesis of old and new material agency.

The act of drawing is a way to visualize and render with visual form; to map and comprehend, to make sense of the structures and laws that bind our world. To draw with a pencil (wood and graphite) is to allow the shaping of conscious thought in various ways – spatially, emotively, schematically, abstractly, diagrammatically. The potential for these modes of expression to be possible is inherent in the material properties of the matter and medium (in this case, impermanence, speed, portability, etcetera). Of course, beneath the surface of this (obvious, old) material agency exists a wealth of hidden or invisible agencies. Graphene is one such layer of agency.

The process through which the material of graphene was extracted was as simple and definitively un-anachronistic as a laboratory experiment, resulting from the use of scotch tape to whittle down graphite to a single sheet of graphene. These experiments, authored by Andre Geim and Konstantin Novoselov at the University of Manchester, resulted in the discovery of the graphene sheet in 2004. It is at this moment in 2004, one of many new-material ruptures of the twenty-first century, when inherited assumptions about an object and medium as ubiquitous as the graphite pencil were tested, and new territories and material possibilities were opened up from underneath. So now, the active agency of the everyday pencil as carbon – with its recently revealed graphene structure – has taken on surreal new forms that one could not have conceived before this moment in time, revealing hidden agencies within the pencil.

The widespread use of graphite in wooden pencils was itself developed from a much earlier discovery – in 1500 CE, large deposits of carbon were found breaching the topsoil in England. The unique mark-making property of the then newly discovered graphite was first used for military purposes,[7] and the pencil, with its distinctive wooden sleeve, was in circulation by the 1560s. Unlike inks and paints, graphite gave impermanency to its user in the act of mark-making. It was this impermanency that became an essential creative agency for artists, scientists and writers – used to make and communicate imperceptible ideas about the world while maintaining a latent sense of probability and erasure.

Graphite's unique agency of impermanence revolutionized modes of expression that appealed to visualization, imaging and recording. Now, the physical material of graphene (carbon atomic sheets) are being tested and probed by the very material agencies and analytical capabilities that the discovery of graphite set in motion in the sixteenth century. Graphite/graphene is now taking on a new material role and agency in the development of cutting edge communication technologies, nanotubes and quantum computers. Therefore, the graphite/graphene moment is a striking example of hidden agencies within materials being unlocked, in effect, by the agencies (and application) of that material in the first place. The transformation of graphite into graphene and its potential to reveal even 'deeper' agencies makes us not only re-evaluate long-held assumptions about the capabilities hidden in pencils, but also challenge underlying presumptions of atomic certainty.

Despite its common usage, graphite is an underexplored medium in the history of drawing, in terms of its use as a specific vehicle for representing images of reality and the imagination. Unlike paint, graphite is rarely explored as medium per se; the agency of the pencil or graphite is so functional that it is often subsumed or rendered invisible in its material sense by its function. While graphite qua substance has been a great enabler of such projects, with its latent ability to articulate and record thought, the critical eye has rarely turned inwards to explore graphite itself. The ubiquity of graphite even in today's digital age attests to the fundamental nature of our relationship with this underexplored, but not underused, medium.

The uses to which we put graphite today are as varied as the scope of human enterprise – from marking sheep like the Borrowdale shepherds who first put graphite to use, to projecting architectural structures, to the nano-extraction of graphene. Graphite remains a quintessential material in mediating the way we perceive the world. I aim to highlight the changing role of graphite in contemporary society with the intention of connecting the relationship between its new and old material agency to changes in our perceptions of reality. In a strange twist of material-driven research, Max Planck was asked by the German Bureau of Statistics in the 1890s to study how to make the most efficient and effective light bulb. Planck's research into what might be considered a fairly routine problem ended up being the discovery of the whole field of quantum mechanics. This story links to the discovery of the 'wonder material' of graphene, as one of the first products to be developed using graphene was a light bulb (2015). The materiality that is intrinsic eventually to the discovery of quantum mechanics is, by its very nature, when charged a materiality that emits light.

The history of drawing details the trajectory of a series of technologies fundamental in the development of human perception. The paraphernalia that permits drawing as practice serves to describe and illustrate our perception, yet also demands an iterative feedback of recognition of the mark itself in the process of making. The pencil-as-medium acts through our hands as an impression-making mark – it allows for gestures that immediately, but with characteristic impermanence, function as a conduit of thought. The pencil has served historically as an essential tool by and through which we have come to make sense of the physical world we inhabit.

The concept of the pencil is relative to the very act of thinking; graphite utilized in a wooden sheath is both conduit and forge for thought that re-examines transference in mark-making. As Fiona Russell writes, graphite in the form of the pencil is 'not only a medium of the artist's ideas, but a part of the thinking process itself' (2012: 35). Hence the reason why we take up the discussion of graphite – and new technological/scientific developments made in recent years – as an inextricable tool in the conception, expression and reception of ideas. In this discussion, the moment of the physical discovery of graphene stands out as a catalyst and disruption, and serves to challenge, by extension, all mediums that we as humans utilize as an extension of consciousness and that, by virtue of their material characteristic, shape the temporality of our world view. This instrumental moment of the discovery of graphene enables us to explore the new material developments and research within the domain of drawing.

The haptic Atomic Force Microscope

As technologies, which serve as extensions of consciousness in our attempts to articulate and map our understanding, graphite and the pencil serve as focal points in the technical development of thought-mapping. From Richard Feynman's pencilled diagrams, graphene, and nanotubes to the instrumental mechanic Atomic Force Microscope (AFM) cantilever, there is an analogous historical link to the process of drawing through a plethora of mediums and technologies. Scientific research, which we might characterize as an exploratory recording and instrumentalizing of the world via the AFM's pencil-like probe, is of particular interest in drawing from natural systems. The pencil tip, in effect, re-enacts (or pre-enacts) those scientific instrumentational inquiries that function in the same way the old gramophone needle does when it picks up the variations in the grooves of a record.

As the hand touches the prosthetic pencil in an act of transmission, it draws on the phenomenological acts of touching and being touched (Fig. 29). If you are trying to convey affection or even love through your touch to another person, then the way you go about touching is the 'proof' of this affection. The emotion that you are attempting to transmit is more than just a moment of recognition. Similarly, making marks constitutes a whole range of emotions, senses and activities within the body that flow out through the prosthetic devices we touch and through which we extend our touch. The receptive agency of paper, as the recipient of the graphic touch, is fundamentally constructed from the cellulose that makes up our own organic bodies.

From a physics perspective – and at the molecular level – touch is an 'electromagnetic interaction'. When we are putting pressure on the surface of our skin, our atoms are themselves applying electromagnetic force onto the atomic surface of the object of our touch, which responds with its own electromagnetic repulsion. There is no, and is never, actual and phenomenal touch. We are physically repulsive and isolated and bound up by electromagnetic energy, but at the same time we are as boundary-less and indefinite as physical molecular structures. If we are not affecting touch on the outside, then our internal organs and systems are not touching but interacting through electromagnetic repulsion.

The context of touch is developed in a paper by Barad, titled 'On touching – The inhuman that therefore I am', who offers a reading of touch as instrumental in what has been touched. The act of measurement, in the sense aforementioned, is also a form of touching, as are chemical reactions. 'Maxwell's demon, like every good experimentalist, would be lost if he or she did not have a highly-developed sense of touch; a feel for the instruments and molecules which constitute our physical interaction with the material world ("Good hands",

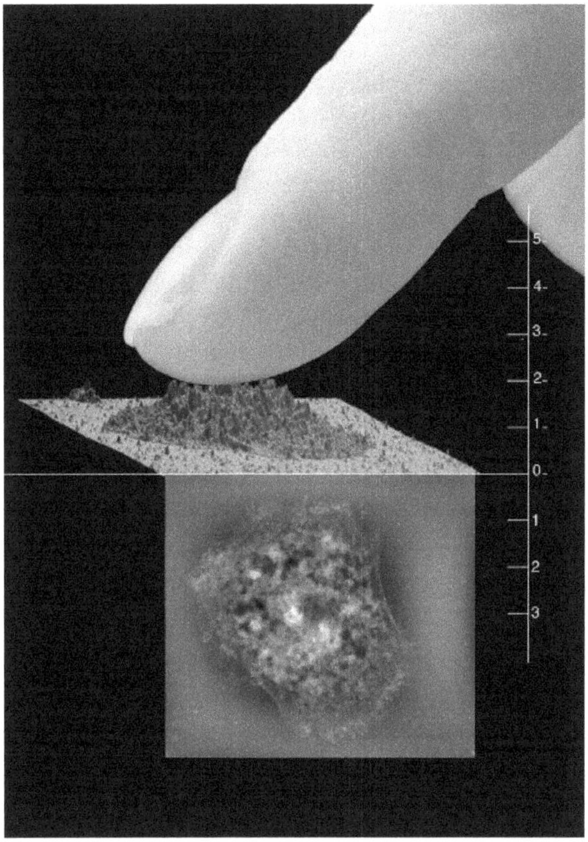

Figure 29: Paul Thomas, *Midas Touch.* Photograph construction for the cover of *Leonardo Journal,* 2:3. Image by Paul Thomas, 2009.

that's what it's called)' (Barad 2012b: 208). Touch grants us a haptic sense of causality, whether we generally acknowledge it or not, whatever we may think of this charged and highly important term. Touch moves and affects what it effects.

The act of touch in drawing with a pencil is a shift between electromagnetic waves wherein the touch of the instrument (the pencil) reacts, through repulsion, to its wooden casing, which in turn responds to electromagnetic resonances in the graphite. The circuit is completed when the graphite touches the paper and its signal is thus transmitted. Shifts and minute disturbances are indexed in this transmission via variations in temperature, emotion, intent and atmospheric disturbances when the transmission is taking place; all environmental factors affect the mark in some way. Seeing the pencil as a prosthetic probe, and the graphene within as an essential element and conduit in transmission repulsion and reflection, reveals aspects of the pencil's latent qualities of agency.

A current fundamental change in our conscious understanding of materiality has, in part, evolved through the adoption of a quantifiable machinic understanding of the world. Through the instruments of nanotechnology, a shift has occurred in the visualization of matter at an atomic scale. We now see evidence made visible in the imagery produced by these haptic instrumental tools of nanotechnology, such as the Scanning Tunnelling Microscope (STM) (1981) and later the Atomic Force Microscope (AFM) (1986). The AFM is at the forefront of imaging technology, able to measure and manipulate matter at mind-bogglingly smaller scales. These are the vanguard tools for mapping the atomic world.

Optical microscopes were used exclusively until 1931 when the first electron microscope was introduced. Optical microscopes have a number of inherent limitations. For example, optical microscopes can only image dark objects or objects with strong refractive characteristics effectively. The diffraction limits resolution to approximately 0.2 micrometres and ambient light can diffuse the focus. Unlike previous microscopes, the AFM is not optical and does not suffer from its technical limitations. The AFM is one in and of a general category of Scanning Probe Microscopes (SPMs) that includes both the STM and the AFM, and its name is in fact a misnomer.

The AFM uses a cantilever probe with a pyramid shaped tip measuring approximately 10 nanometres (approximately 100 atoms). The tip shown in Fig. 30 is attracted to surfaces in a scanning process. An image of the surface is obtained by the microscope mechanically moving the probe in a raster scan of the specimen, line by line, and recording the probe–surface interaction via a laser beam reflected from the tip to a photodiode. Applying here the concept of the gathering of scientific data through a process of touch, the haptic interface of the AFM emulates the phenomenon of touching. The AFM's 'feeling' of substrate surfaces effectively anticipates and prompts a reconceptualization of the yet-dominant ocular-centric processes of understanding and engaging with the world. The AFM and nano-assembly/nano-fabrication 'feel' the potential for moving atoms around and spur the possibilities of creating new connections and, therefore, new materials, forms and structures. This process of construction by the rapid prototyping-printing of atoms thereby replicates the biblical origin story of the construction of humans from dust particles (Thomas 2008).

In the article 'Nanotech, blur and tragedy in recent artworks by Gerhard Richter', Kristian Hvidtfelt Nielsen defines a particular context to frame the works of Richter. Richter challenges technological and scientific developments as being the cornerstone of the modern world that 'have frequently been turned against civilization' (2008: 487). A series of prints reference Richter's use of images from an STM and AFM as backgrounds. The images differ from the 'ontology of ordinary vision', in that the AFM, in contact mode and raster scan, creates by default a drag-like blur of the process (Nielsen 2008: 487). It is as though you were touching an atomic surface with your finger, feeling your way across the surface, smoothing, smudging, compressing, erasing and altering the very surface being recorded. Richter blurs the blur of the atomic world by evidencing this very smudging of the atomic world. This is shown in his work *Graphit* (2005), a silkscreen over an offset lithograph image from an AFM scan of graphite. The underwhelming visual spectacle of regimented rows of atoms (an aspiration of science to control the world atom by atom) is in stark contrast to the uncertainty stated by Heisenberg's uncertainty principle.

The blur of the AFM imaging gives us the instrumental image (not to be confused with optics), not based on sight but on touch resistance, attraction and repulsion. What the scientists and the software engineers are exploring is the possibility to be able to measure proximity and therefore write out the glitches. In Richter's work, the structure of materiality is contrasted with the unpredictable nature of paint, the 'rational visual practices of nanotech and the irregularity and contingency of art' (Nielsen 2008: 488).

The impenetrable blur that is found in Richter's work has an indexical link to the quantum smear, where things can't be revealed in definite locations and the movement of time can't be recorded. The blurred/smeared image stops you from moving through and into the image, rather it creates a layer that appears as though one is walking in a dense fog. The more you try and define an object, the more lost and paranoid you become; the blurred image is a

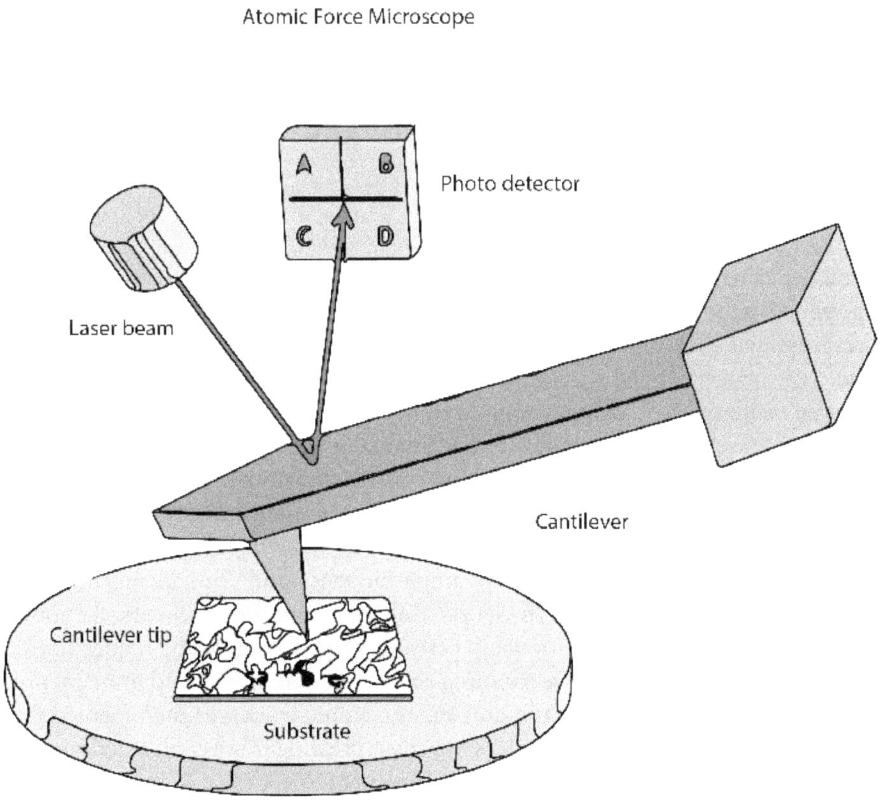

Figure 30: Diagram of an Atomic Force Microscope cantilever scanning process. The diagram shows the cantilever tip used for the precise raster scanning of atoms from the sample substrate. Graphic by Paul Thomas, 2018.

perceptual doorway to chaos. Richter demonstrates how we cannot conceive of dealing with a world that has no definite focus or boundaries.

Scientists, when using technologies such as the AFM (Fig. 30), feel their way in the dark, using the indications of their touching and probing of the subatomic world. This scientific exploration in total darkness is what will generate the alchemical future. Scientists fiddling in the dark beg the question: how do the creative arts, the visual arts, respond to science that is happening in total darkness, not in the shadows? The darkness is a continuation of the void. Graphite, when used to put marks on paper, creates a grey black that, when rubbed with a cloth, starts to shine. This graphite shine lures one into the richness of the medium, and creates and permits the simple transference through the phenomenal touch of the carbon agent.

The humble pencil is a symbolic holdover for the expression of intellect. Moving beyond the world of immediate (artistic) expression, the pencil, for all its simplicity, is history's dynamo: the medium through which we have sought to make sense of our terrestrial existence. Plans, taxonomies and diagrams – the process of drawing is based on the particular properties inherent in the materiality of the stylus: charcoal, contè and now graphite. A drawing does not simply articulate some cognitive schema, it constructs a sense of reality as much as it attempts to elucidate a pre-existing one. The art of drawing is a discursive act in which the pencil, as a tool, is mediating between the 'seer' and the seen.

Agency and mediums

The practice of drawing from the medium of graphite enables the user to rethink the marks being applied to the surface in terms of ideas that can be instantly reflected upon, manipulated and re-presented back into the world. The nature of graphite is that the weak atomic bonds that hold the material together, when applied to a surface, make a continuous dry mark that can also be easily erased (a counterpoint to stone, marble, monumentality and permanence). The ability to erase the mark instantly is instrumental in being able to rethink plans, diagrams, landscapes and so forth, as the work is being constructed. Robert Rauschenberg, who spent time erasing his own drawings as a creative process, wanted to test this concept out more fully through an external subject. *Erased de Kooning Drawing* (1953) (Figs. 31 and 32) by Rauschenberg is a graphic example of the impermanency and permanency of graphite even in the esteemed hand of the artist. In George Cotkins' chapter on the work, he points to how Rauschenberg challenges the relationship between artist and object by requesting a drawing by Willem de Kooning to erase. De Kooning chose a work that included more mediums than just graphite, such as ink, grease and crayon. The iconic work was then formed from more permanent mediums that were not so easy to erase, creating more contradictions inherent in the whole material and conceptual process of erasure (2015).

In the impermanency of the pencil drawing, like the impermanency of life, is the nature of the medium's demonstrable vitalistic carbon-based agency. Unlike Malevich's *Black Square* (1915) where a previous painting was covered by a layer of black paint, graphite layering is

Figure 31: Robert Rauschenberg, *Erased de Kooning Drawing*, 1953. Traces of drawing media on paper with label and gilded frame; 64.14cm x 55.25cm x 1.27cm); San Francisco Museum of Modern Art. Purchase through a gift of Phyllis C. Wallis; © Robert Rauschenberg Foundation. Licensed by VAGA, New York; photograph courtsey of Ben Blackwell.

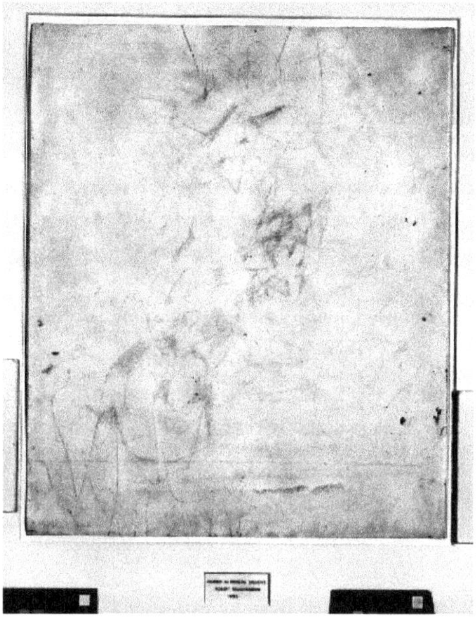

Figure 32: Robert Rauschenberg, *Erased de Kooning Drawing* (digitally enhanced infrared image), 1953. Traces of drawing media on paper with label and gilded frame; 64.14cm x 55.25cm x 1.27cm); San Francisco Museum of Modern Art. Purchase through a gift of Phyllis C. Wallis; © Robert Rauschenberg Foundation. Licensed by VAGA, New York; photograph courtsey of Ben Blackwell.

qualitatively different now. Thus, graphite is not only useful in its latent ability to make marks but also in its ability to be erased (Figs. 31 and 32). With Rauschenberg, we have the agency of the pencil mark being bestowed with the significance of the prominence of de Kooning, only to show that its value is transitory. The process of drawing is by default thinking on paper. It is not merely to draw out ideas or to represent things, but rather drawing helps us to see, comprehend and to sense more holistically. The historic discovery of graphene enables us to now rethink and change the way we view the material world and therefore the power of the medium's ability to respond to the way that we represent the world.

Malevich's *Black Square* is the epitome of deletion of an image, as opposed to its erasure. The blackest black (trade name Vantablack) is now in the province of graphite not paint. Single sheets of graphene rolled into nanotubes and stacked together are responsible for capturing light. Frederik de Wilde's art shows that graphene has the ability not to overpaint the image, but absorb all light to deny the image a chance of being transmitted. The medium inherently draws 99.96 per cent of light into its atomic body. Currently, the artist Anish Kapoor, who in 2016 gained exclusive rights to Vantablack, is unironically constituting a new version of Rauschenberg's erasure of de Kooning. The artist's work no longer needs to be erased as the material is in itself an erasure of light. Therefore, any image that is transmitted is absorbed in the nanotube's molecular materiality.

De Wilde's work *Hostage* (2010) developed in collaboration with NASA and Professor Pulickel Ajayan, Robert Vajtai and Daniel Hashim from the Rice University chemistry lab explores the aesthetic and conceptual realm of the blackest black (Thomas 2013). The medium graphite, which illuminates and makes visible imaginings, holds a dark secret that de Wilde tested with *Hostage*. By owning the rights to the blackest black, Kapoor attempts to erase and extinguish the evidence of the existence of all other works, whereas de Wilde's work *NASABlck-Crcl#1* (2013) (Fig. 33) exemplifies the need for material agency to be free for artists to explore. Duchamp's exploration of a mode of collapsing the artwork in *Boîte-en-valise* (1935–41) can be seen in the singularity of de Wilde's conceptual work *NASABlck-Crcl#1* that posits all artworks as visually being absorbed and captured in material's blackness.

Contemporary physics, with its use of probability and uncertainty, has affected the way artists perceive and interpret the world. The impact of probability and uncertainty was initially challenged and explored by the Futurists and Dadaists at the beginning of the

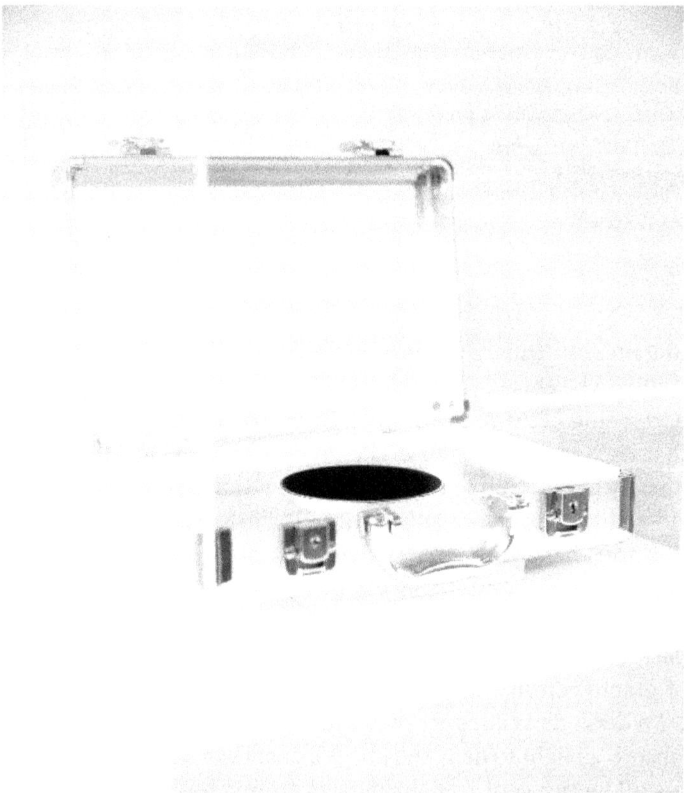

Figure 33: Frederik De Wilde, *NASABlck-Crcl#1*, 2013. © Frederik De Wilde.

twentieth century. The simple word 'hold' when examined in the context of contemporary physics becomes more significant as we continually contest the 'hold' or grasp we have on reality, that things still 'hold' true in what we know as the classical world, as opposed to the quantum world. The classical world, where things have fixed positions, differs from the quantum world where it is impossible to measure things with any certainty. Here in the classical world, a 'hold' (like a ship's hold) can be seen as a container of truths to be shifted from space to space, unseen in their containment. It is the hold that contains us, that we sink into, becoming the ruination of our ability to imagine. The hold is not a grasp but rather a strangling grip on reality that denies all other probable realities for the sake of a monoculture. The concept of the 'hold' humans have on reality needs to be adjusted, shaken and ruptured to enable a release from the classical world to a comprehension of the quantum world.

To unpack Delayed Choice Quantum Erasure (DCQE), we need to cite the famous 'double slit' experiment and the role of the observer.[8] How can an artist measure the world if the intention of the artist as the observer is conceptually already decided? In physics, the very act that shifts the pattern of our reality is based on the object, and the agency of observations. The observation makes the object perform differently by refreshing its identity under the continual scrutiny of the significant gaze. In quantum mechanics, a shift in the position of the observer is created in the DCQE experiment. This experiment is demonstrated in the diagram of the 'double slit' experiment that reveals the standard interference pattern when not measured (Fig. 34a). With DCQE, if you were to observe the waves after having gone through the double slit and are just about to hit a sensor forming a wave interference pattern at that very moment of being observed, the waves transform into particles (Fig. 34b). These particles would appear to instantly travel back in time through the double slit to re-enact the performance, but this time as particles with no wave-like interference pattern appearing on the sensor (Fig. 34c). This experiment is based on when the observation is made at a particular point in time, to make the particles reveal which slit they had moved through, but instead the particles retrace their steps going back in time to act as particles and therefore produce no interference patterns but a standard cluster pattern on the screen.

The DCQE suggests that the observer is instrumental in the outcome, demonstrating an ontological understanding of the world. The act of measured observation creates a rupture in a form of erasure that makes all the world disappear, to reform as what could be seen as a refreshed or remediated picture of the classical known world. The erasure experiment suggests alternative narratives based on concepts of atemporality and the untimely.

The recurring abstraction of reality in an artwork is based on the intensity of observation, as related to that which is observed. The artist is confronted by time that appears to travel backwards and reform as a new present overriding the previous present. The artist becomes a co-conspirator in DCQE as each observation affects how the object world is being seen. In this exploration of DCQE, the interference pattern, we want to take the act of the artist drawing or painting as an equivalent ongoing re-enactment of the phenomena of DCQE.

To explore the analogous relationship between DCQE and art, it is useful to consider how the artist looks at the world and makes a mark on a surface, as to what he/she really sees. The very act of observing changes what is observed and therefore the first mark must change as its meaning is erased by the second mark. It is a continual process of adjustments that can replicate a smearing on the surface.[9] The observed world appears to comprehend the attempt to be measured by adjusting the information being transmitted to the observer. This mirrors the DCQE where the artist as measurer attempts to comprehend what is being

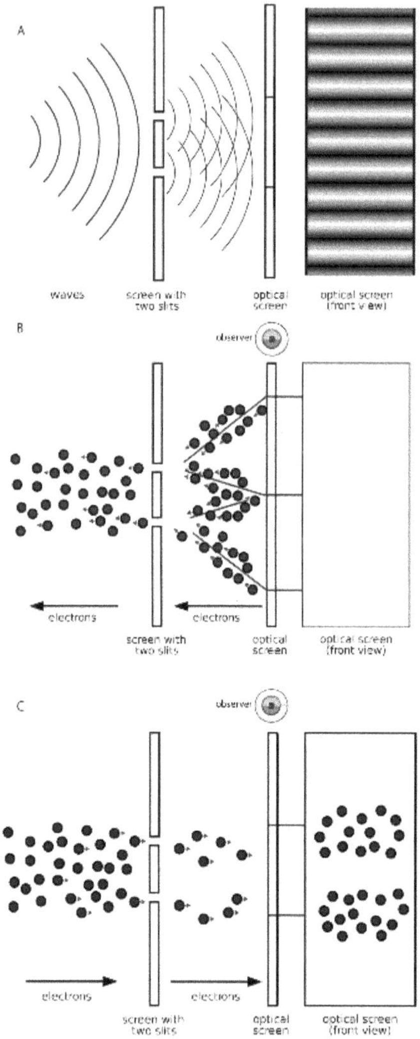

Figure 34: Quantum erasure diagram. (A) Standard interference pattern from the double slit experiment. (B) To try and capture which slit the wave/particles pass through, the observation happens before it hits the screen. Once observed the wave is transferred to particles and moves back through the double slits. (C) The particles then go back through the slits to form a cluster. Graphic by Paul Thomas, 2018.

measured. The artist, therefore, measures through the act of observation, as a wave becomes a particle. Realizing it has been seen, the particle goes back through time to re-enact the process as purely a classical description.

This going back in time is what I am referring to in the development of a work of art that seriously contests what the world is. It is the artist's choice to measure that has an influence on what is being seen and understood. The adjustment is done by the information being transmitted, reversing and going back to the object to be retransmitted to the artist-observer.

Now we have time going in reverse, enabling a new interaction with the world to take place. This space becomes the atemporal liminal space between the known and the unknown. This observation is not the simple act of seeing or a simple recognition of what is presented, but a performative measurement of the world.

The critical gaze of the serious artist is one that sets the DCQE in motion. The time shifts back and forth, erasing not only the real time but also challenging the very marks that were once truthful as being now false markers of reality. The never-ending agony of trying to capture what cannot be captured, to see that the critical gaze makes the world adjust, reflects the failure of the drawn mark to hold any reality of the world. This failure is the beauty, courage, sensitivity and intensity of an artist's endeavour, in continually attempting to capture a truth born of intuition and adversity.

The graphene experience was explored in 2015 by Cornelia Parker, a well-known artist who was asked to produce a work for the opening of the new Whitworth Art Gallery at Manchester University. The university's Nobel Prize winner, Konstantin Novoselov, was introduced to Parker as a potential collaborator to develop the work for the new gallery.

Their interest in the wonder material of graphene focused around some eminent drawings from the Whitworth collection where their intention was to harvest some minute samples of the graphene. From amongst the works, Parker chose William Blake's *Study for Tiriel Denouncing his Sons and Daughters* (1789). Novoselov extracted samples of graphite that were sub-100 microns (0.1mm) in scale that he reduced down to a single sheet of graphene.

The potential of the poetics for new agencies residing in the infinitesimal element of graphene extracted from the drawing was the very essence of Blake's touch. The data could have explored more fully at a nano level what the graphene quantitatively or qualitatively could reveal. The sample of graphene was used by Parker as an interface to a poetic elucidation of a Blake poem in the form of a firework display (Youngs 2015). In *Breath of a Physicist*, Novoselov created a sensor from the graphene sample that would trigger the firework display (Fig. 35). The atomic erasure takes the erasure of de Kooning to extremes of scale, and therefore intent, of what is revealed by this action. How much can you extract from an artwork before it becomes a new artwork?

What Richter, de Wilde and Parker point to are aspects of nanotechnology and quantum phenomena that are emerging onto the radar of contemporary culture.

Graphene, the name for a single atomic sheet of graphite (a pure form of carbon), is an excellent conductor of electricity and is fit for various mechanical purposes. Since its discovery in 2004, graphene has been applied to a great many new purposes – from the construction

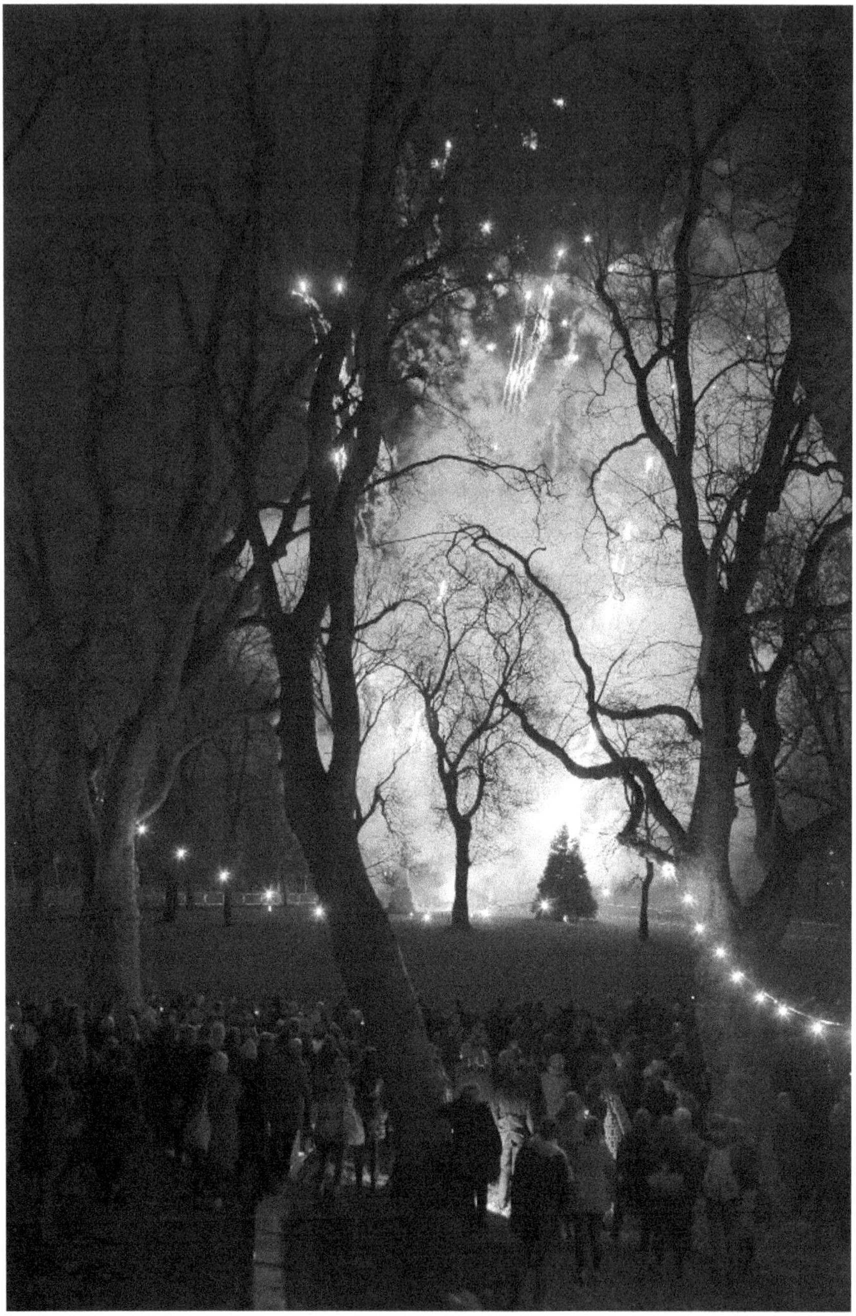

Figure 35: Cornelia Parker's *Blakean Abstract* opening night at the Whitworth – Fireworks with a meteor inside (set off by Breath of a Physicist). Photograph courtesy of David Levene.

of next-generation transistors to light-weight, high-strength composite materials, flexible television screens and even quantum computing. The potential of this substrate of graphite is being explored across many university research facilities. Among these, the Centre for Quantum Computation (CQC) at the University of Oxford led by theoretical physicist Dr Simon Benjamin in the area of 'Molecular & Solid-State' is of particular note. Current quantum computer research at Oxford theorizes the use of graphene in the form of single-sheet nanotubes, as well as the nano-buckyballs, that have the potential to advance the study of quantum computation (Benjamin et al. 2006). Benjamin's current theories of controlling spin in the uncontrollable superposition are fundamental to quantum computer research, where flux, oscillation and vibration are at the very atomic centre of its universe. The quantum spin is based on controlling qubits as a foundational utilization of materiality.[10]

The research into quantum computing and the spin is dealt with in depth in the Spin chapter. There I considered the use of graphite (carbon) to create a dialogue, attempting to demonstrate and expand upon the involvement of the materiality and agency of graphene at the core of the new 'medium' of quantum computing. What we are now presented with are the implications and evidences that the discovery of graphene has on the medium of drawing and its subsequent technological legacy. A range of contemporary artists, some mentioned in this chapter, have taken up a revised approach to the materiality of graphite in this age of new materialism.[11] This shift of agency in the potential or perceived materiality of graphite has opened up a rich new environment for interdisciplinary creativity.

Graphite drawing, a medium entrenched in the history of art and visual communication, has prototypically been employed for its surety of expression. Now with the change of agency with the discovery of graphene, the discussion moves to one of fluidity, connectivity, absorbing light, material strength and memory whilst maintaining its impermanence in mark-making. The uses for graphene are growing daily; it is one of the most promising nanomaterials and all this affects the agency of graphite. Graphite offers the artist or mark-maker the possibility of steady, unbroken line-work. For this reason, graphite has become synonymous with both functional diagrammatic schemas and precise anatomical and taxonomical recording. A constructed conceptual framework that includes the latency of graphene underpins the ways in which technological growth impacts the way we see and think about the world.

The art of graphite

The role of graphite in drawing, from Georges-Pierre Seurat to Joyce Hinterding, can be seen as having exposed other meanings that challenge the nature of the material. In Seurat's drawings, we see the graphite conté as a medium that is used to rub over the paper's surface texture with no defining lines, enabling an atmospheric effect to take place (Fig. 36). Conté is a medium developed in 1795 by Nicolas-Jacques Conté to compensate for the global shortage of graphite. Conté succeeded in mixing wax with graphite powder to form a square

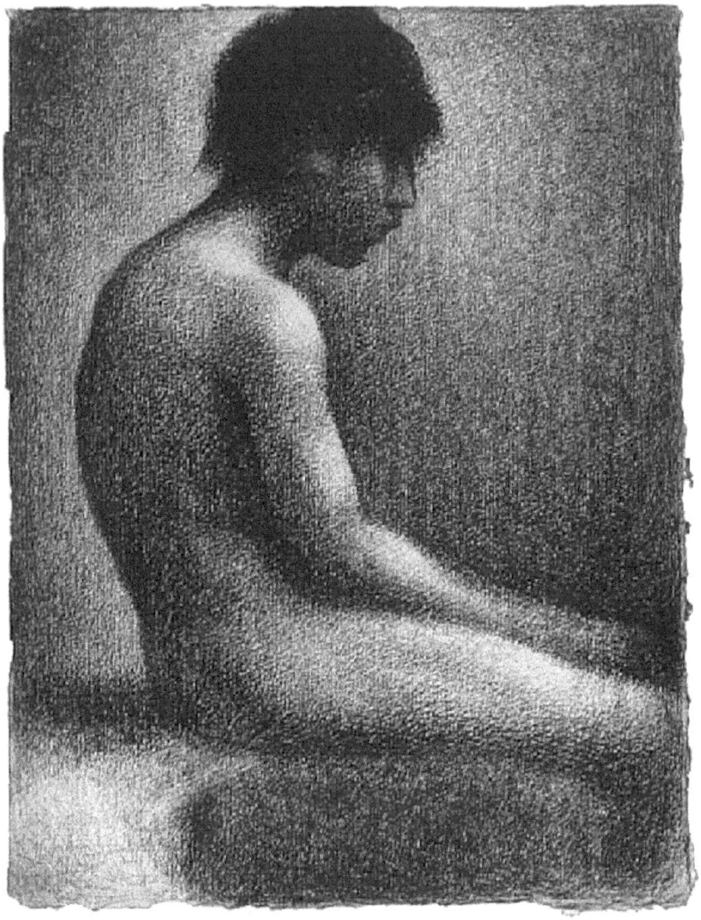

Figure 36: Georges-Pierre Seurat, *Seated Nude: Study for 'Une Baignade'*, 1883. Conté crayon on Michallet paper. Courtesy of National Gallery of Scotland.

stick. The conté-image emerges in conjunction with the texture and material of the Michallet paper, to form a becoming of and in the object world. Here, the world slowly emerges from chaos, through the artist's use of the medium, into a tentative ordered existence to be constructed only by an observer's gaze.

These marks describe a process of materialization, as recognizable shapes evolve through gestures and touch to form cloud-like swarms of graphite. The outward expression of the tonality of the graphite medium evokes forms of a becoming – Seurat's marks conjure boundary-less recognitions, as recognizable instances in the process of a world come into existence. The edges blur and are absorbed into the background as the paper surface texture gives a sense of a striated space.

In contrast to Seurat, Hinterding's use of graphite exposes its latent potential as a medium that is at one with its powers for visualization and its capabilities for electromagnetic transmission. Graphite is deposited as a confluent layer on the paper's surface in a series of symbolic forms indexical to a circuit diagram (Fig. 37). The symbolic forms are precisely constructed to look at the medium as a means of transmitting and receiving electrical signals (an example of artists using graphite as a conduit for signals).

Hinterding's work draws on aspects of the invisible spectrum of waves that are brought to life through the drawing as antennas,[12] transmitting through graphite to receptors that amplify the signal as a sonic experience (Fig. 38). The translation of signals via the powers of the graphene (at the core of graphite) demonstrates the beginning of a different understanding of the material or drawing. These elements of the electromagnetic spectrum are made poetically visible in Hinterding's work and expose the act of drawing as a direct relationship with the potential of a graphic antenna. Here, the white sheet of paper with a thin two-centimetre black pencil line creates a visual formal labyrinth. The graphite on the sheets of paper, in their desired form, is reminiscent of old square antennae recording universal waves of information independently from the role of the artwork. Hinterding's graphite/graphene experiments are effectively recontextualizing the materiality of graphite as a conduit/conductor for transmission and reception of electromagnetic resonances.

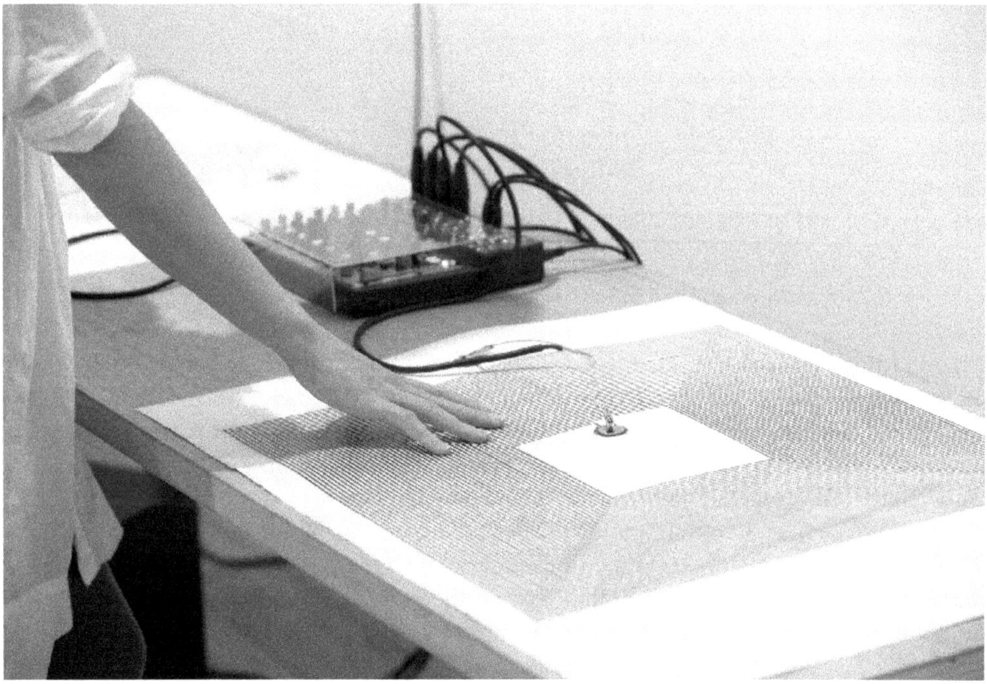

Figure 37: Joyce Hinterding, *Aura*, 2009. Photograph by David Haines. Courtesy of the artist and Sarah Cottier Gallery.

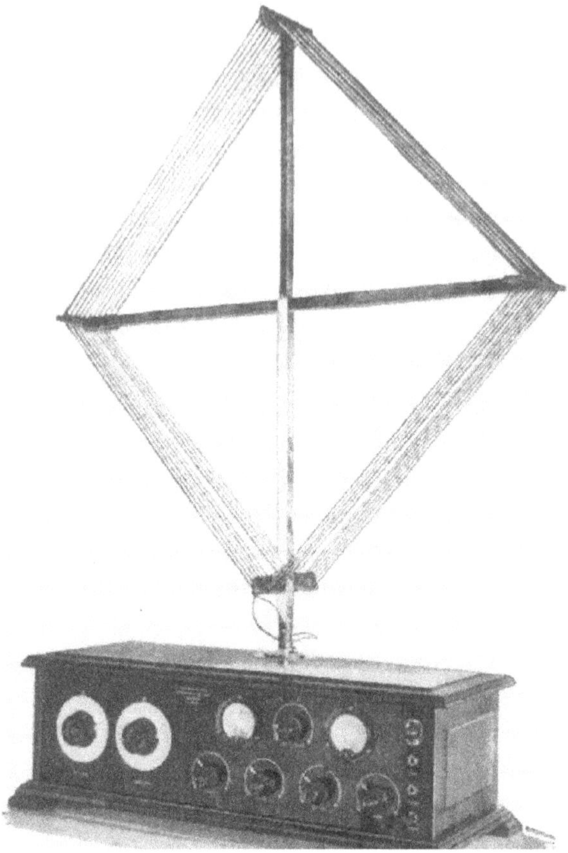

Figure 38: Early radio with external loop antenna. Loop antennas have the ability to eliminate some electrical interference compared with wire antennas. Loop antenna reacts to the magnetic portion and ignores the electrical component, where the noise is concentrated. Photograph in public domain.

Hinterding's artwork *The Oscillators* (2003) develops a connection between the erased de Kooning drawing and Frederik de Wilde's *NASABlck-Crcl#1* (2013). *The Oscillators* installation consists of three sound-producing drawings, each created with graphite and based on an electrical oscillation diagram, a complete work within a work. The oscillators are activated from a gallery light via a solar panel. The electricity is then fed through silver leaf capacitors where the configuration creates a feedback oscillation so the work feeds back through itself. In an interview, Hinterding explains how the electricity appeared to learn its way in the artwork through the graphite when the material conditions changed from the studio to the gallery. The audible pitch started to increase in the artwork, towards moving out of hearing range. The graphite drawing of the capacitor needed to be erased so the work could stay in an audible range (2016). The artwork was then subjected to ongoing adjustments of erasing the graphite from drawings through its malleable properties. This process, which took place during the exhibition in the gallery, challenges the drawing's integrity and the authenticity of its marks, to ensure that the totality of the sonification is maintained. The 'pencil lines are both a representative gesture and a layer of thin conductive stone' (DeLys et al. 1996), enabling the relationships between all the performative material agents.

Hinterding explains her intentions as being framed by a question of what happens when an artist tries to visualize the invisible forces within and of the cosmos into a work of art. Hinterding states that a 'kind of challenge takes place: the challenge to see what we might call the synthetic world as part of what we might call the natural world' (DeLys et al. 1996). On her website, she states her intentions for the work *Simple Forces* (2013):

> A spiral of graphite becomes a resonating structure that emanates energy, scavenging, storing and – when amplified – making audible and 'playable' the atmospheric forces in the room. The geometric forms are taken from algorithmic structures from the parallel worlds of nature and science. Symbols – a feature of art from its very beginnings – also carry the hidden agency of a functioning circuit, with the ability to relay 'extra territorial forces' or to 'sing the space'. Far from poetic fictions these works function: constantly energised with the electromagnetic, the circuits are active and constantly flowing.
>
> (Hinterding 2012)

The antenna is a useful metaphor when dealing with this topic, as listening devices pick up signals as they relate to the potential of the body to receive more information than it would appear to know how to deal with – broadening the breadth of the potential reception of signals (Fig. 39). In 1964, the background microwave radiation, residual of the Big Bang, was discovered by Arno Allan Penzias and Robert Woodrow Wilson as they experimented with the Holmdel Horn Antenna (Fig. 40). This particular antenna (discarded by Bell Lab) was an unusual looking device shaped like a megaphone. Penzias and Wilson attempted to discover information about microwaves at low temperatures, initially by using this antenna. In the data they kept receiving, there was a lot of noise, and no matter what they pointed the antenna at or how much they tried to clean and stabilize the equipment, the noise continued. After initially anticipating that pigeons were the culprit, they deduced that this was cosmic microwave background radiation raining down from outside our galaxy.

The functional symbolic forms of Hinterding's images expand the material awareness we have of graphite and its properties to transmit and receive energy. Furthermore, and by extension, Hinterding puts us in touch with the 18 per cent carbon in our own bodies and the potential for comprehending the transmissions internally in the human body from the energy being received. The energy from the electromagnetic spectrum that we as mobile human antennas are able to receive seems limitless. The potential is in finding a way for our intuition to trust our own translations of some of these signals. Artists finely tuning into the everyday signals that bombard our bodies, and then translating these transmissions to be represented as a work of art, demand a trust in their intuition that enables new visual languages to emerge.

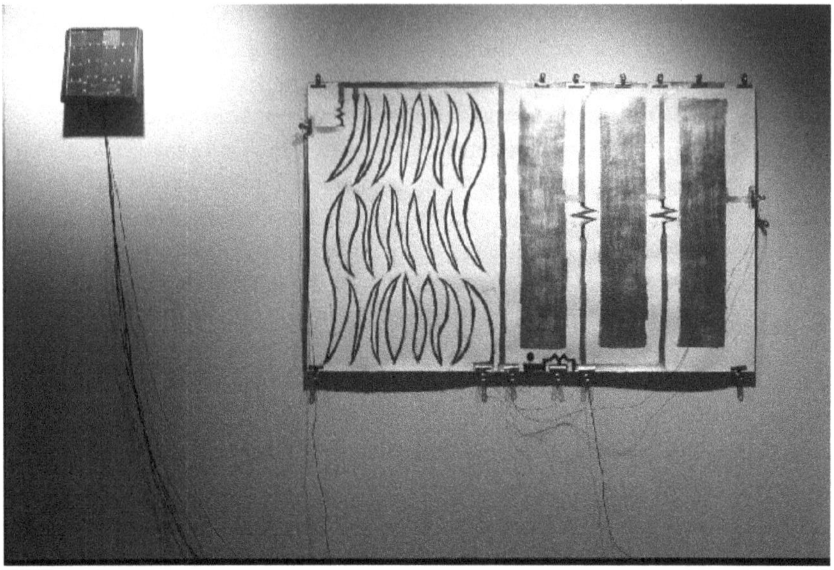

Figure 39: Joyce Hinterding, *The Oscillators*, 1995. Photographer: Ian Hobbs. Courtesy of the artist and Sarah Cottier Gallery.

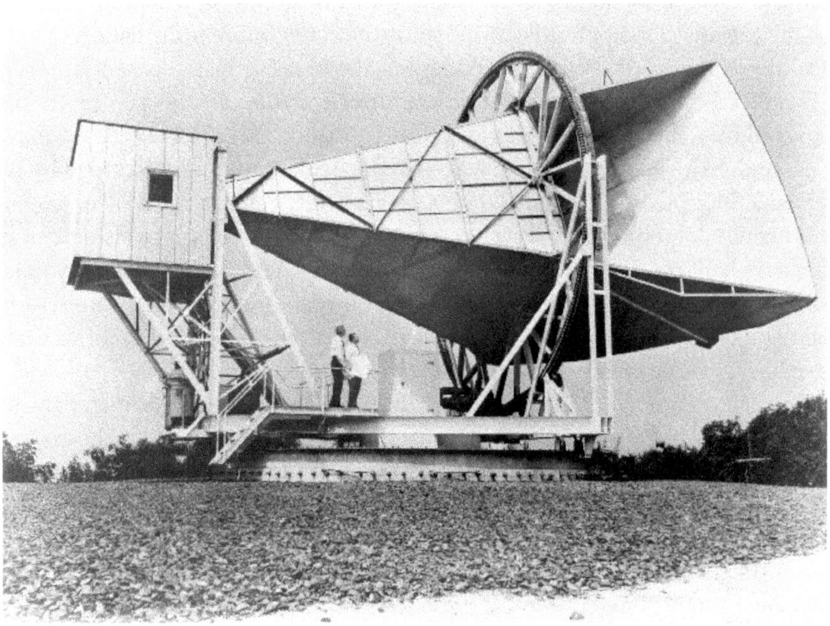

Figure 40: *Holmdel Horn Antenna,* The 15-metre antenna at Bell Telephone Laboratories in Holmdel, New Jersey was built in 1959. Photograph in public domain.

This challenge to our understanding of the materiality/agency of graphene is played out in another form by Victoria Vesna's (artist) and James Gimzewski's (physicist) artwork *Zerowave Functions* (2002) (Thomas 2013: 19). These sculpted buckyballs are interacted with by the audience's shadows – the artists' way of imaging the inner-agency of its atomistic media specificity. In doing so, these artists take the viewer inside the atomic world and enable users to draw and shape the invisible. The interaction with the molecule C60 forms complex relationships with matter emulating a playful control by the viewer but opens up the questions of ownership, domination, manipulation and colonization of the material world. From Seurat's marks, gridded and granularized by the surface mixed with the gesture, emerge Lucretian atomistic molecules, then matter and finally nature. The process of exploring the graphene material's performative agency, that is now to be put to use in the construction of Benjamin's quantum computing theoretical research, demonstrates the development of our ongoing relationships in the micro world with the immaterial, while engaging in the undetectable possibilities of quantum thought.

Drawing out

In this act, drawing as a form of measuring is part of the history of graphite: from the initial discovery of graphite deposits under a tree at Borrowdale in the United Kingdom to the newfound technological significance of this fundamental art-making material. The consequences that the revelation of graphene will now have on the practice and epistemology of drawing are of great interest. The art historical importance for researching the emerging and manifold potential materiality of graphite is to outline the how, why and what of a radical reinvigoration of the medium.

That the second most common element in the human body is carbon (averaging 18 per cent, after oxygen at 65 per cent) draws interesting parallels with the alchemical genealogy of the pencil. What could we make of the relationship between alchemy and mark-making in artistic practice, chemical components of material graphite and that of the human body? Do we start to see these attributes as being causal, intrinsically linked to being human? What connections can we identify between a consciousness latent in matter and the actual material that makes the invisible world visible? Synergies coagulate around thinking as being atomic, the use of carbon atoms in the processing power of the quantum computer and the potential of graphene as a telecommunication tool. These synergies lead to discussion about what sort of matter/material/materialism provides the basic agency of consciousness, allowing it to enter into the world. Such possibilities, at the very least, prompt a rethinking of the way we currently hold as significant those marks we have made and their histories.

New materialism reveals the invisible possibilities and latent agencies of matter – revealing matter per se – as much more powerful than we could have initially imagined in expressing human thought (and the sedimentary records of time and space that it stores like data). Graphite may now serve not only to record and map thoughts, but to inscribe a range of non-semiotic utilities, such as its aptitude as conductor of electromagnetic currents. Graphite-cum-pencil has been the defining material of the modern age in visualizing key components in the developments of science and industry – engineering, architecture, etcetera. Now, with the discovery of graphene, the very stuff that allowed for the conceptual exploration and mapping of key scientific developments is itself being unlocked as a fundamentally significant product in the new age of scientific discovery.

The agency of graphene, latent within graphite, takes carbon from out of its conceptual/mapping schema into a new realm of high-tech computer/material agency where it is necessary for quantum computing nano-technology and so forth. This chapter has articulated the role of graphite/graphene as a material for communication in the dawn of the quantum age. What role can the medium play in the hands of an artist who is attempting to deal with phenomena at an infinitely small scale?

The transforming agency of graphite and the definition for new modalities of drawing are central to graphene as a material in the production of contemporary monitors and touch screens, and are of great significance to the interests of new materialism. The confluent material agency of graphite as tool for visualizing the world both physically and now digitally earmarks the possibility of a new kind of drawing. The same material that has historically been used by artists is central to mainstream digital mediums of communication today. The reestablishment of touch as the sense of connecting directly with the materiality of being in the world is highlighted by the transmissions made through the pencil to the surface of the paper. As mentioned, the needle of the 'feeling' AFM becomes a seismic oscillator: a catalyst to translate energies that are drawn out from the undulations of the record of atomic grooves and which are then translated into sound and image. This re-engineering and re-imaging of material information (from their natural sources) – such as marks of graphite on the paper surface – can be seen to be gaining insight into the intentionality exhumed from the mark-making of the artist. The archaeological archivist's comprehension of the history of the pencil and its marks realizes that there is more to contribute to the knowledge of society by examination of the very nature of the medium.

The inauguration of graphene has prompted the need for a revisionary history and reconceptualization of graphite drawing and material agency more broadly. In light of the new agencies now detectable and latent in graphite, can drawing remain isolated from its contemporary 'reading'? These new and experimental interpretations of old materials show the need for expanding sensory awareness of our surroundings and how we interrelate and interact in them.

Performative agency

The performative[12] agency of materials that go beyond a human intent is a major factor in the discussion of graphite and the pencil, as it is used in the construction and enabling of expressions of meaning in and of the world. Here, we can consider the performative agency of materials (new materialism) in relation to the resonant (and relative) energies thereof, at the atomic level. If we are to produce some relationship between the material world and our macro world, we as humans need to understand the reciprocal performative agency that exists at scales beyond/below our immediate perception in the material mediums we use to comprehend it (hence reciprocal). Between the domain and artifice of the Anthropocene and the organic natural systems of the non-human cosmos is a common mediating relationship of matter where the interplay between electrons and neutrons communicate across the various forces that bind material (atomic forces).

Latent material agencies are enhanced or unlocked by processes aimed at (re-) comprehending performativity as a quality inherent in matter. By noting the role of context in regards to material agency we can perceive new possibilities of application, that is, new possibilities of what to do with the layered and latent agency of materials. We might call this the contextual enrichment of material agency. The experiments that seek to 'enrich' material agency are speculative in nature because, by definition, the parameters of experimentation cannot be fixed. However, as we have explored, these revelations of new agencies in materials can be progressively unlocked by technological innovation, and as such open up the discussions of new materialism to the co-conspiratorial relationship between matter, human and technè.

Why do specific tools arising from the discovery of new materials become positive forces for growth whereas some tools become an obstacle to progress and development linked to humanity's downfall? The reciprocal relationship between human innovation, technology and material agency is organized or structured in such a way that in 'unlocking' latent agencies within materials (like carbon) we may consequentially (for better or worse) produce new technological tools and utilities that subsequently allow us to dig deeper into the latent agency of materials. Alongside graphite, we have its carbon 'cousin' coal and carbon 'sister' diamond – three manifestations of carbon.

How much of the 18 per cent of our own organic bodies formed from carbon is relevant in terms of our capacity for consciousness? In other words, how much of our carbon-selves are involved not only in raw consciousness, but also in the tool-using capabilities of cognition that subsequently and historically have utilized the agencies of carbon we find in nature? How much carbon is consciously appropriating its own agency?

When it comes to material choices for technological implements that contribute to human existence through extending our material-sensory sensitivity, how much of our chemical makeup is directly constituted and appropriated in conscious behaviour/choices? How can we see this performativity taking place in the material world, which is now being made visible through the imaging science technologies? In *Matter and Memory*, written at the dawn of the twentieth century, Henri Bergson writes:

> Matter thus resolves itself into numberless vibrations, all linked together in uninterrupted continuity, all bound up with each other, and travelling in every direction like shivers through an immense body. In short, try first to connect together the discontinuous objects of daily experience; then resolve the motionless continuity of their qualities into

> vibrations on the spot; finally fix your attention on these movements, by abstracting from the divisible space which underlies them and considering only their mobility (that undivided act which our consciousness becomes aware of in our own movements): you will thus obtain a vision of matter, fatiguing perhaps for your imagination, but pure, and freed from all that the exigencies of life compel you to add to it in external perception.
> (Bergson 1990: 276–77)

The uninterrupted continuity of the material-atomic world is created from infinite vibrations that conjoin and thread through all matter and tether the human and non-human into a singular material-cosmic organism. This total vibratory mass is the base materialism of consciousness – and yet at the same time it is the object-oriented ontology that we have forever sought to comprehend and explain as object, other-than-self. The vibrations of the cosmos are not separated/separate from us via the frequency of waves; rather, the universe-as-vibration is the very network of stimulus and tensions that interconnect us with and as the world. Lived experience is one that is part of a continuum of material vibrations – albeit a category and agency of materialism that is invisible to the naked eye. We cannot know this by privileging the visible portion of the entire spectrum of electromagnetic resonances above all other energy transmissions to and from the material body. We must overcome the selective monopoly of sight and the resonances of visible light to enable ourselves to break free from the limitations it puts on our perceptual comprehension of the material world.

We have covered the relationship between graphite and graphene to discover something about the agential relationship that exists between materials and human interventions of creative mark-making. Once we perceive what was at first invisible (and also inconceivable), new complexities emerge and our understanding and expression of the material world shifts into a new paradigm/model.

We understand from the Epicureans that thinking and thoughts are atomic in nature (and thus material) and stem from an intuitive understanding of the necessity of there being something (as opposed to nothing) as the base structure of the cosmos. Nothing cannot be proven to exist (that is, there is no proof of nothing); therefore, everything is something in the material sense – including thought – and as the smallest individual agent of the material world is the atom, which is the stuff of all matter, thoughts must also be atomic (that is, the new materialism of the quantum age extends to operations of thought).

The writings of the Epicurean Lucretius, in his poem 'Of the Nature of Things', led to the conception of the atomic swerve as the material phenomenon that generates, through the clashes borne of its directional agency, the entirety of (classical) matter (Carus 1950). The Clinamen or 'atomic swerve', material oscillation, vibration and/or atomic axial 'spin' of the particles are all connected. By extension, the agency of atomic swerve is also the heart of quantum computational capability. In this lineage – by placing the genealogy of thought as rooted in atomic processes – the fundamental fiction at stake in the perceived separateness of the thinker and the thought is revealed. Thinking the world into existence becomes part of the process of humanity's latent desire to comprehend and 'know' the universe and our role within it. In reality, as we have suggested, it is a desire to know ourselves that drives us out into the material world.

Therefore, in the twenty-first century, man is on the cusp of creating quantum thinking machines (computers) generated by the same atoms utilized in the material processes (agency) of our own consciousness. This development demonstrates what we already really intuitively understand; what some artists have sought to show through their practice and integration of materials for centuries. The point–line–plane of Klee's paintings are endemic to understanding these primal actions at a quantum level – even though he never knew the quantum mechanical world, he intuited the condition. The point of mark-making for Klee was seen as being indivisible in the void, the stable atom, the problematic stability/instability from which atomic direction (swerve) is used to build and construct other worlds. Klee intuited that the static point is not the smallest point, that the point is not a static condition, that the permutations of the point are like the fluctuations in the vacuum energy field.

The graphite point of the pencil (a unique material agency of carbon) beats with the (hidden) agency of a graphene heart. Graphene, as we have seen, is, by virtue of it being composed of carbon, a major constituent of the human-organic body. It is also a material agency that is contributing to quantum computing research and therefore to the agency of thought that is inherent in the non-static, static point. Klee writes:

> The pictorial symbol for this 'non-concept' is the point that is really not a point, the mathematical point. The nowhere-existent something or the somewhere-existent nothing is a non-conceptual concept of freedom from opposition. If we express it in terms of the perceptible (as though drawing up a balance sheet of chaos), we arrive at the concept grey, at the fateful point between coming-into-being and passing-away: the grey point. The point is grey because it is neither white nor black or because it is white and black at the same time.
>
> (Klee 1961: 3)

Buried in notes he penned in the early 1920s, Klee mulled over the concept of 'the point that is really not a point' (1961: 3). In the same way, fifteen years later Schrödinger (1935) (Trimmer 1980) conceptualized the quantum cat in his favourite analogy as occupying both two polar states and yet none of them at once (Trimmer 1980); this was Schrödinger's cat as the 'grey point' of quantum materialism – that 'fateful point between coming-into-being and passing-away' – a cat (impossibly) both alive and dead at the same time.

The grey point is an inexplicable point that marks the oscillation between up and down. This oscillation gives us a 'point' of becoming with the 'non-conceptual concept' of the quantum problem and theory of superposition. The non-dimensional is the expression of the probability of discovering an undiscoverable position. Klee expresses concepts or non-concepts of chaos in a similar way in his visual renderings of the greyness of an impossible point/non-point. This expression follows on from the observational reasoning of attempting to interpret what constitutes our understanding and model of reality.

The grey point is therefore a particle that must also be a wave; a quantum intuition that reveals the magic of Klee's perception and his foresight of what is truly taking place at the quantum core of the materiality of nature. The whole passage of Klee's reads like a poetic premonition of the reality and problem of quantum superposition. Klee's work, in light of his musings, serves to analogize the material universe (including our own consciousness) as a unified fabric whereby the slightest change (swerve) or material disturbance brings the world, or part of it, into existence. Deleuze was fascinated by Klee's articulation of the grey point, stating:

> Paul Klee presented these three aspects, and their interlinkage, in a most profound way. He calls the black hole a 'gray point' for pictorial reasons. The gray point starts out as nonlocalizable, nondimensional chaos, the force of chaos, a tangled bundle of aberrant lines. Then the point 'jumps over itself' and radiates a dimensional space with horizontal layers, vertical cross sections, unwritten customary lines, a whole terrestrial interior force (this force also appears, at a more relaxed pace, in the atmosphere and in water). The gray point (black hole) has thus jumped from one state to another, and no longer represents chaos but the abode or home. Finally, the point launches out of itself, impelled by wandering centrifugal forces that fan out to the sphere of the cosmos.
>
> (Deleuze and Guattari 1987: 312)

Deleuze was also interested in the philosophical concept of expressive inflection, about which he writes the following:

> Inflection is the ideal genetic element of the variable curve or fold. Inflection is the authentic atom, the elastic point. That is what Klee extracts as the genetic element of the active, spontaneous line [...] For Klee, however, the point as a 'nonconceptual concept of noncontradiction' moves along an inflection. It is the point of inflection itself, where the tangent crosses the curve.
>
> (Deleuze 1993: 14)

'Inflection', in this context, is where vector points-as-atoms become lines, as their visual traces are mapped as falling in the void. Hereby one point-as-atom unexpectedly swerves and so becomes the spontaneous line, the intuitive line. Where the spontaneous, unpredictable swerve makes contact, it creates emergences and generates the becoming of matter. In this Deleuzian reframing, inflection thus becomes the phenomenal agency of 'expressive matter' wherein the fabric of the universe itself is intuited and expressed so that it might be seen and felt.

The material by-product of the history of mankind's objectivity and intervention has produced very real threats to the stability of our material ecology. However, it seems improbable that further human interventions will solve the waste-product and mess of our own existence. We must explore the substantial change that needs to occur in humans, rather than doubtful Band-Aid measures aimed at finding a panacea. The narcissism of the Anthropocene is total; we cannot undo ourselves. Is there a rupture that can occur in the human psyche that will enable us to see the atomic truth of our ongoing coexistence with the material world, of which we are constituted but still treat as other? Being, as we are, constituted of the same material weave that constitutes the cosmos, the perspective we have cultivated that looks to the universe as an object distinct from us is problematic at best. This anthropocentric gaze, which has allowed us rational contemplation, is the essential technical factor that has spurred our cognitive distancing from the world; rendering it 'object' when (in reality) it is a co-constituted subject. The anthropocentrism of the Euclidian gaze is one that seeks to measure the universe in such a fashion that each individual is prevented from embracing the world-as-self, that is, insufficiently addressed (and can't be rewoven) by existing terminology. The shift that needs to take place for this break to be restored follows on from a pursuit begun in twentieth-century art, where perspectival space was being reconfigured so that the observer might reconnect with a sense of being part of the material world he/she is observing and not a voyeur of a distant object and 'other'.

In this context, I want to place the agent's role of being as that of the artist, who historically is seen to be well placed as a receiver and transmitter of world experience via visual and immaterial material manifestations, bringing the world and all the sensory signals into poetic human expression that challenges the very nature of what it is to be conscious.

Notes

1 Quantum-oriented ontology is a neologism designed to bring the concepts of quantum phenomena in alignment with the nature of being. Schrödinger's cat thought experiment, live matter–dead matter, superposition (multiple states at the same time), the role and position of the observer and entanglement are brought into this discourse where the influence of quantum mechanics has been on all forms of culture.

2 'Quantum matter' is a term used in this book to describe an immaterial thingness. That is, the idea of a thing in the process of forming, half idea, half form. The object forming is independent of human comprehension of the object's development.

3 I am coining the term 'quantum-oriented ontology' to reflect that quantum underpins the very fundamental nature of all objects.

4 Agential as performer is borrowed from Karen Barad's agential realism that suggests that 'it privileges neither the material nor the cultural: the apparatus of bodily production is material-cultural, and so is agential reality'. While theoretical constructs are not to be understood as representing transparently given observation-independent properties possessed by independent material objects/beings as they exist in isolation from all observational interactions, neither are we to interpret these constructs as artefacts of the observational process, purely discursive gestures imprinted on the blank slate of passive matter (Barad 1996: 180–81).

5 Intra-action is another neologism from Barad that draws on an intra (from within) to create relationships between the individual and material: ‘In other words, measurements of the values of the well-defined variables are attributable to the phenomenon as a particular instance of wholeness, the fully contextual be-in where the matter and meaning meet’ (Barad 1996: 179).

6 In 1803, John Dalton (1766–1844), an English chemist, instigated a study of atomic theory. He became the first scientist to explain the behaviour of atoms in terms of the measurement of weight. He also uncovered the fact that atoms couldn’t be created or destroyed. He situated the atom as the building block of all matter based on Democritus’s observations made during the classical Greek era.

7 Graphite was a good lubricant for allowing cannonballs to be fired cleanly. Graphite was the material used to line moulds for cannonballs in 1553; therefore, the supplies were protected by the state.

8 The interference pattern (Fig. 34a) demonstrates that light acts as waves when travelling through the double slit and when measured acts as particles.

9 Quantum smearing means that electrons that can’t be measured, acting like clouds, can move temporally backwards and forwards – like a classical smear can move a drawn line back and forth as though it has travelled back in time as well as the future, with the approximate position being the darkest point of the line, a form of interference pattern.

10 ‘Thus, one possible building block of the quantum computer is the buckyball C60. The C60 molecules have individual nitrogen atoms fired at them that become trapped inside the buckyball, creating nC60 molecule, which is at the very core of a microprocessor’s array. The caged nitrogen atom within the buckyball is incarcerated and isolated but still has the ability to spin and can be controlled by microwaves to emulate the zeros (by spinning up) and ones (by spinning down) of the digital computer’s binary code. However, quantum computing explores the potential of the nitrogen atom’s superposition inside the buckyball to be in more than one state (not zero or one) at the same time’ (Thomas 2013: 17).

11 *Graphite* was a group exhibition held in October 2012 at GV Art, London, curated by Robert Devcic, http://www.gvart.co.uk/graphite-exhibition-group-show-october-2012.html (Collis 2012).

12 Hinterding was inspired by Loop Antennas and has built a number of them as one of her seminal starting points as sculptural pieces (Hinterding 2016).

13 The performative expression that is linked through art ‘is precisely a contestation of the excessive power granted to language to determine what is real’ (Barad 2007: 133).

Chapter 5

Cloud

The itinerant cloud

The broad concepts of atmospherics are analogous to the shifting molecular world. The analogy supports the inclusion of art and culture in the realm of molecular fields and the quantum phenomena that is manifest in the invisible void.

It is not difficult to conceive of the molecular world as constitutive of the kinds of fluctuations and variations – which we know intuitively at the scale of objects – as atmospherics. This analogy aids in the imaging of an inclusive field of artistic and creative culture (not purely scientific) in terms of the interpenetrations between quantum-atomic fields and human intellection and inquiry. We have brought together these realms and their interactions across history, and qualify this expanded field of human inquiry into the physical world as cloud-ness. It is the cloud that best serves to analogize the uncertainness that characterizes contemporary science, creativity and art.

The first moment of the emergence of the uncertainty of cloud-ness, one of two historical moments discussed here, occurred in the 1420s. The art historical cloud was encountered and problematized as non-perspectival space by Filippo Brunelleschi who needed to resolve perspectival problems in order to complete his peephole device. This device was designed to eliminate the uncertainness and non-perspectival character of those literal clouds that populated atmospheric space. Brunelleschi's success in revealing a unique quality of the cloud was inadvertent because his characterization of them as chaotic was due to the clouds interrupting his goal of constructing a perspectival, architectural understanding and image of the world. I want to establish a link between the way the cloud has been explicitly marginalized by the doctrine of perspectivalism that determines the way the world is perceived and current understandings of the data cloud. This section instigates the analysis of material agency but with a focus on how technologies define and shape world views. It explores how the immateriality of data creates a simile to the physical cloud in conforming to invisible/visible manifestations of reality.

The second pivotal moment of the emergence of the cloud-as-world-view extends into the explicit metaphor of cloud computing (1994).

Here we compare and explore the cloud as revitalized non-perspectival world-model that extends as both a ubiquitous system of data and as a molecular archive and distribution process. This revitalization has opened up a potential playground of chaos and data for the interplay of uncertainty and reality to be explored. By using the concept of particles, or bits of data, playing in the metaphorical molecular cloud (Fig. 41), we can apply an experimental theoretical methodology[1] to break down the distinctive boundaries between our own embodied materiality, the object world and the irrational structure of clouds. Thereby, the

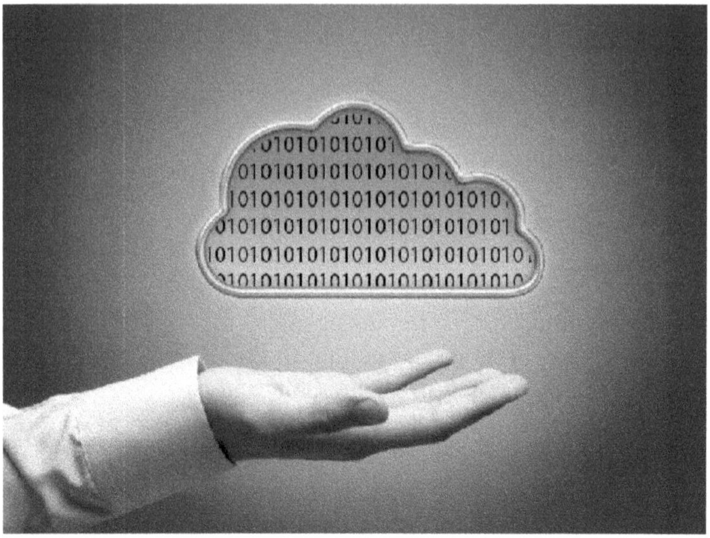

Figure 41: Graphic example of the contemporary corporate cloud iconography. Image by Paul Thomas.

molecular cloud becomes a new atmospheric formation for rethinking data and molecular agency as a global, and seemingly immaterial, distribution of storage and retrieval – of breathing in and breathing out through data.

The molecular aesthetics of the data-chaos cloud is a series of swirls, eddies, vortices; a structure and form that can exist everywhere and yet cannot be said to exist in any one place specifically. This particular gateway pulls all the molecular elements together as a singular hyperobject,[2] similar to the musings of Wil McCarthy's self-replicating nanomachines (called 'Mycora')[3] in the novel *Bloom*.[4]

As with protocols of bit torrent files, the cloud provides a new concept of boundary-less states of assembly, distinct from and beyond the materialist or machinic diagrams,[5] and more integrated than the (precursory) practices of collage and re-mixing or remediation that has become characteristic of late twentieth-century and millennial aesthetics. The cloud is not an object but an experience, and its particles are the very building blocks of a molecular aesthetic in which we live, breathe and act.

Shifting power of perspective

In the doctoral thesis 'Reconfiguring Space' I previously wrote a brief description of Brunelleschi's peephole device that draws from Antonio Manetti's first-hand experience (Thomas 2006). In his biography of Brunelleschi, Manetti enables us to comprehend the role that Brunelleschi's optical device has had in establishing linear perspective and, subsequently, how

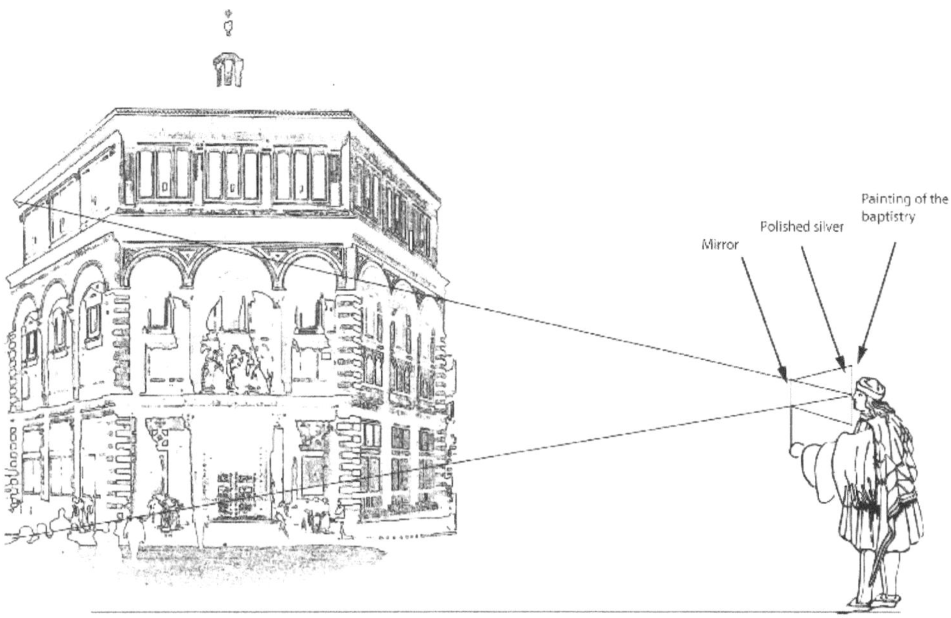

Figure 42: Illustration of Brunelleschi's peephole device pointed at the baptistery of San Giovanni from the doorway of the Cattedrale di Santa Maria del Fiore (Duomo) in Florence. Graphic by Paul Thomas, 2018.

the irrationality of clouds and their reluctance to conform to perspectivalism have been actively suppressed and historically relegated.[6] In this machine, the user/viewer, via a mirrored image of the baptistery painting, sees the assimilation of the view from the doorway of the Duomo (Fig. 42). The viewer peers through a hole in the back of the painting held in one hand. In front of the painting, a mirror is held at arms-length in the other hand. The mirror, which is placed in front of the actual baptistery, creates a seamless image in its reflection with the visible architectural surrounding space flowing on from the edge of the mirror in the hand of the viewer.

The other major technical feature of Brunelleschi's peephole device (Fig. 42) was the clever use of burnished silver to deal with the sky that was omitted from the painted panel, as Manetti writes:

> And he placed burnished silver where the sky had to be represented, that is to say, where the buildings of the painting were free in the air, so that the real air and atmosphere were reflected in it, and thus the clouds seen in the silver are carried along by the wind as it blows.
> (Manetti 1970: 44)

For Brunelleschi, clouds were not seen as being part of the agenda for the invention of perspective; we are left to assume that, as a mere reflection, they were superfluous to the construction of true perspectival space. The conceptual cognition of perspective was not a technical issue, as the cultural audiences in Europe pre-Renaissance did not have any precedent, let alone comprehension, of perspectivalism and its doctrine. As Brunelleschi's scopophilic view of the world was naturalized, clouds grew into a symbol of resistance for anti-perspectivalism.

Erwin Panofsky writes on this topic to give an insight into the profound effect that this shift away from an experiential understanding of space had:

> In a sense, perspective transforms psychophysiological space into mathematical space. It negates the differences between front and back, between right and left, between bodies and intervening space ('empty' space), so that the sum of all the parts of space and all its contents are absorbed into a single 'quantum continuum.'[7] It forgets that we see not with a single fixed eye but with two constantly moving eyes, resulting in spheroidal field of vision. It takes no account of the enormous difference between the psychologically conditioned 'visual image' through which the visible world is brought to our consciousness, and the mechanically conditioned 'retinal image' which paints itself upon our physical eye.
>
> (Panofsky 1991: 31)

The cloud as un-perspectivalized territory, as a rogue element in perspectival order, is now being undermined within a system of top-down re-colonization. Here we see the libertarian symbolism of the clouds stripped of their independence. The cloud is colonized and reunited within the perspectival schema of power and authority. In the twenty-first century, all future-planning is digitalized and relayed in a top-down fashion so that Brunelleschi's divine perspectival order, the eye of God, is undone, but only to be replaced by large subterranean data storage facilities iconized as the cloud. People do not need to plan ahead; planning is instantaneous and taking place in real time; orders, directions and suggestions are being signalled from above into the mobile phone via various applications now epitomized by the cloud connection. The network/cloud does theocracy better than any government of God.

Images of nature (clouds and sky) seem trapped within the shine of the polished silver. The perspectival moment was a fundamentally significant shift in the European perception of the world. Brunelleschi's work contributed to transforming the way that we see the world by shifting the status of the observation by universalizing a quantifiable apparatus (the peephole device) so that even when it is not physically present, it nonetheless mediates the way in which we see the world – as if we were not just present but omnipresent, seeing through God's eyes.

The moment of not knowing precedes the moment of recognition; it is a fine line between these two states of ignorance and knowledge. The moment of perspectival recognition cannot be reversed. Despite our attempts to escape the perspectival order, like clouds, perspectivalism is how we now see the world – so that, as I have written elsewhere, the 'paradigmatic mathematical theory of single-point perspective is fundamental to the imagining and construction of space in modern times' (Thomas 2006).

The cloud becomes the 'other',[8] the remainder, left out of the equation of perspectivalism that was brought into being, revealingly, through a peephole device. In this context and in this role as resistor, the cloud becomes an animate, autonomous and yet amorphous mass that can herald freedom or serve as the harbinger of gloom and doom. As Damisch writes:

> Brunelleschi's experiment reduced/cloud/ to an effect of reflection, a mirror image engineered within the figurative field by dint of a material artifice. But the significance of that reduction is twofold, both rational and symptomatic: it was at once a consequence and evidence of the institution of perspective space as the theoretical space of (re)presentation.
>
> (Damisch 2002: 183)

Classical architectural traditions serve to mediate and maximize the symbolic and the practical functions of space. In this tradition, architecture sought to mitigate the operations of chance and unpredictability so as to minimize any latent threat to determine architectural boundaries. Yet, the cloud can serve architectural interests in its capacity as a 'negentropic vortex'[9] – a mapped, ordered and corporatized space that deals with ownership, permanency and certainty. This ordering of chaos, which emerged in the digital age, serves as the absolute antithesis to the natural resistance of the cloud to territorialization.

Clouds give us insights into the quantum world of uncertainty of entanglement,[10] between what is experienced in entangled forms of air and water. The cloud then, with its uncertainties and vagaries, becomes an avenue for quantum thought and molecular thinking. In this vein, we can consider the primal sources of life; vapour and moisture intermingling in the air, forced to ascend due to its geographic location and physical properties, where it transfers from invisible to visible in its nebulous state to take the form of a cloud (visible to invisible to visible). These clouds, as they hover over the earth's surface and are observed by painters, are witnessed as nature's symbolically anti-perspectival agents. Those artists who tapped into this symbolism attempted to frame clouds as a counterpoint to rationalist analysis and perspectivalism – not as an effort to map the swirls and vortices therein, for this would be self-defeating, but to simply permit them, allow them their own agency. Thus, clouds accrued a symbolic, anarchic power set against the status-quo politics of authority and control. The challenges are not purely symbolic, however, in that they engage in the history of cultural discourse, and as such, manifest very real challenges to the perspectival order. It is worth quoting McLuhan and Parker here as they define this anti-perspectival antagonism of the cloud:

> The Renaissance was unconsciously engaged in creating a pervasive visual space that was uniform, continuous and connected, but the Middle Ages had had a very different kind of space as its physics and social environment. One of the reasons for this was the 'idea' behind Medieval representation. It was the idea rather than the psychological

> narrative connectives between figures that was central to Medieval communication. Thus there was a no need for a 'rational' or continuous space in which the figures could find psychological interaction. At one Renaissance extreme, the spectator finds it easy to put himself into the painting as if his space and the space in the painting were the same. This became possible when the 'vanishing point' was established 'inside' the painting. With the vanishing point came the illusion that space was a continuum between spectator and art situation. The spectator becomes part of the lines of force which find their focus in the vanishing point.
>
> (McLuhan and Parker 1968: 12)

Before the onset of the European Renaissance, social, religious and political control came inexorably and directly from above. The structure of a pyramid serves as a model, with religion or the state giving orders directly from the top-most point. The European Renaissance vanishing point of power and control shifted to the horizon following the birth of modern perspective. The vanishing point that is made on the horizon line is fictional, an unassailable point of converging lines that in reality never converge and therefore cannot be reached. The promise of perspectivalism is that it offers one the opportunity to visualize oneself as being positioned at an ultimate point, whereas in reality such an achievement is impossible.

The false promises of the autonomy of self that perspectivalism offered are now promised again through the data cloud. Thus, power, information and energy are now all repositioned so that it is hierarchically overhead, and perspective's vanishing point is repositioned to be directly above us (Fig. 43c). The duplicitous incorporation of the metaphorical cloud whilst repositioning perspectival power to above-ness consumed this rogue element and institutionalized it. The individual now has potential around the clock global access to information and social interactivity. Information about contemporary

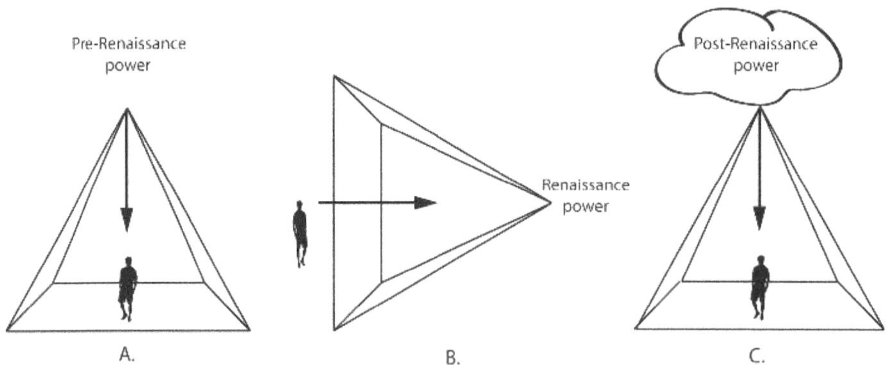

Figure 43: Diagram of perspectival power triangle demonstrating repositioning. (A) Pre-Renaissance power from above. (B) Power is repositioned in the Renaissance to the vanishing point. (C) The Post-Renaissance power has been repositioned to above, incorporating the cloud. Graphic by Paul Thomas, 2018.

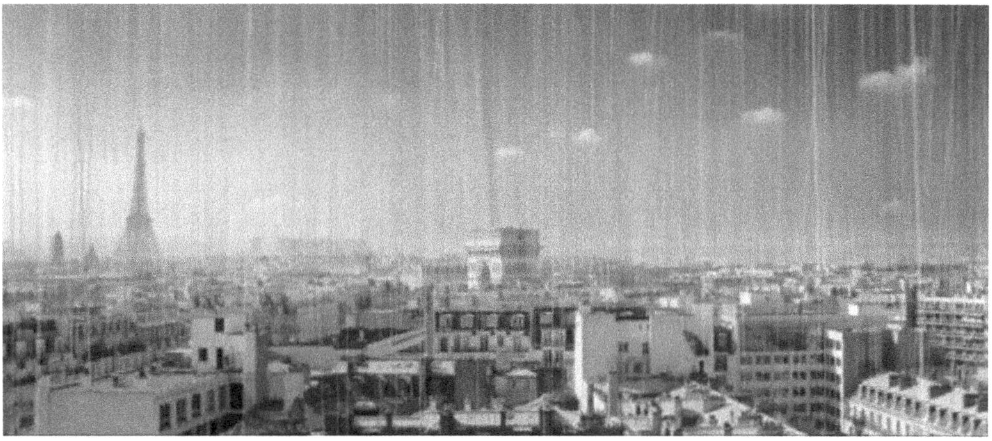

Figure 44: Still from *Lucy* (Luc Besson, 2014), showing a visualization of data coming down from above. Courtesy of EuropaCorp.

daily life can stream down literally all the time, to direct and coerce the user who is never, as such, free from the possibilities of being directed (Fig. 44). This time includes the un-perspectivalized and rogue character of the cloud as we have explored it in its elemental model, incorporating all aspects now of repositioning power back to a new post-perspectival world view.

Pre-cloud conscious

I want to situate the dawn of the data cloud in the work and writing of media art pioneer Roy Ascott, who proposed in his paper 'Nature II' of 1993 that we were entering molecular time (2013). Ascott was instrumental in facilitating an emergence of artists working directly into and with the technologies borne of the network age of telecommunications. I.P. Sharp, a forward-thinking company at that time, provided an electronic exchange network system known as ARTEX, due to the efforts of Adrian X, Bill Bartlett and Gottfried Bach, creating a space for artists working in the 1980s. In many ways, ARTEX was the precursor to the conceptual schema of the data cloud and it was ARTEX that enabled use of the first international artistic text and cursory graphic collaborations over digital networks. This emerging conceptual-network schema preceded and pre-empted the rise of global data and telepresence.

The prescience of visionary works like *Hole in Space* (1980) by Kit Galloway and Sherrie Rabinowitz (Fig. 45) served to demonstrate the slippages that creative practices were experiencing in the transitional period into the age of information and of global networks.[11] *Hole in Space* constituted a real-time satellite uplink between Los Angeles and New York where projections placed in shop windows streamed live in such a way so as to enable people

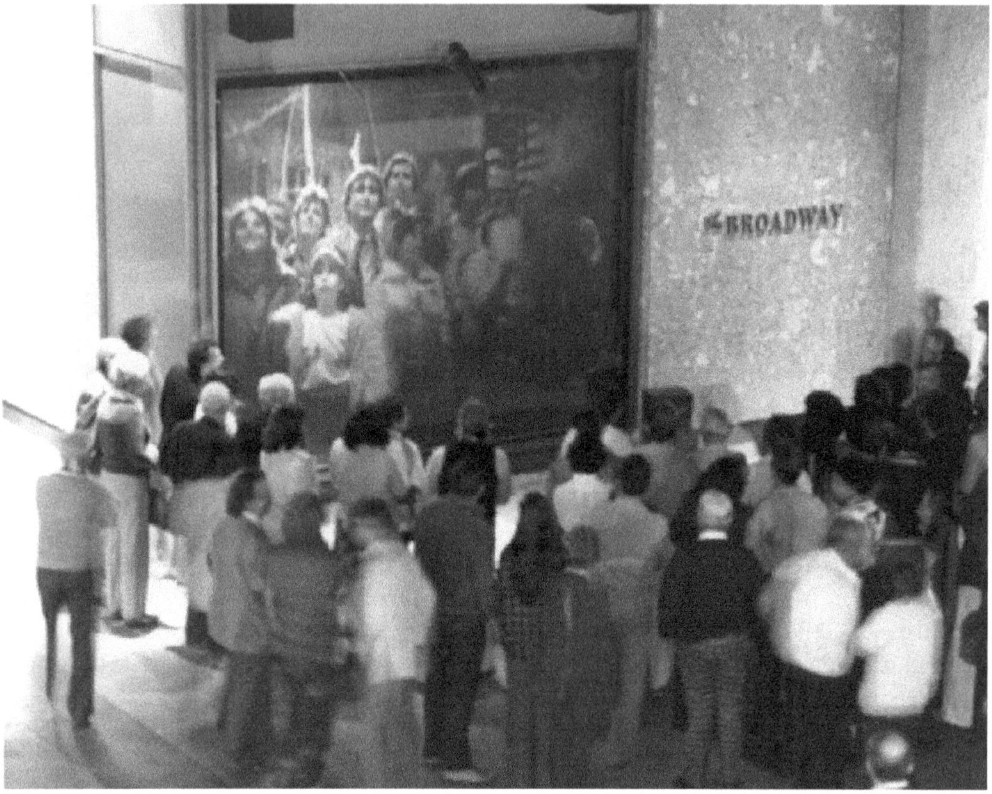

Figure 45: Kit Galloway and Sherrie Rabinowitz, *Hole in Space*, 1980. Still image from video documentation. © Galloway and Rabinowitz, courtesy of artists.

to interact with each other. This extraordinary event created a telepresence in real space and opened up a new way of thinking about the material constitution of virtual space. The work demonstrated a role of entanglement as human interaction, 'corporeal materiality at a distance, physically inaccessible or otherwise remote' (Ascott 1990: 246).

The consequences of the work *Hole in Space* can be seen in the telepresence work of Paul Sermon,[12] whose seminal piece *Telematic Dreaming* (1992) (Fig. 46) captured the imagination of touching, feeling, intimidation and violation at a distance. Two people, each one lying on a separate bed in a separate location, are joined together via the network to appear connected together on each other's bed.[13] The ethical and moral implications of 'touching at a distance' a disembodied, real-time image of a person, draw attention to the proprietary concerns related to telepresence. The concerns pre-empted the boundaryless body as a swirling mass, a digital/molecular augmented entity, that can seamlessly be deterritorialized and reterritorialized.

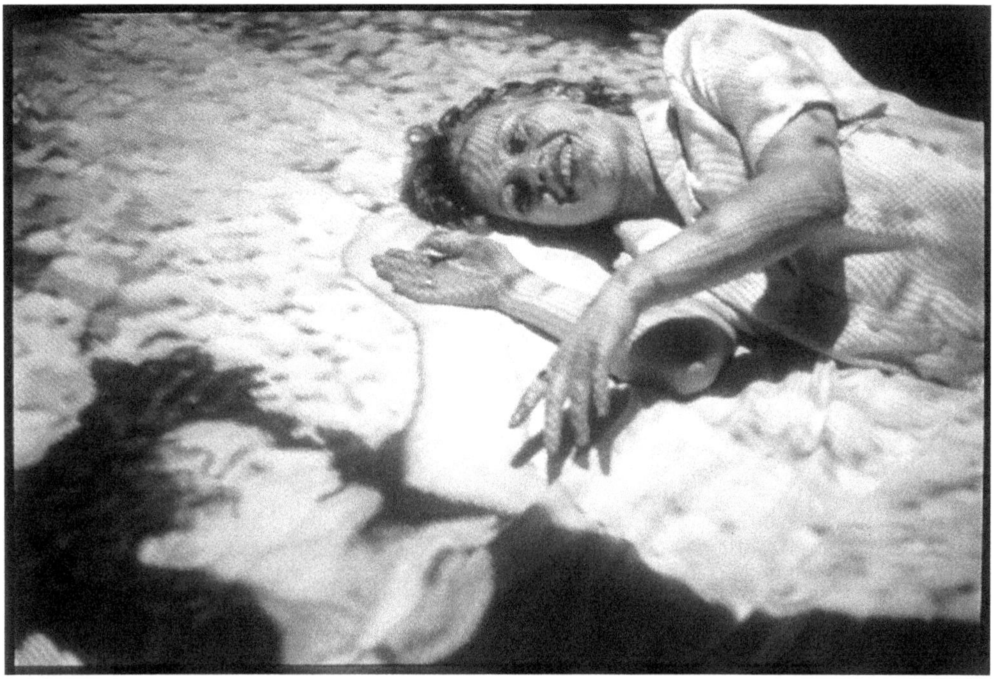

Figure 46: Paul Sermon, *Telematic Dreaming*, 1992. Photo courtesy of the artist.

By the time of *Hole in Space*, Ascott had already conceived of the implications and ramifications of a global network, a vision that informed his early network projects (Fig. 47). As Ascott himself wrote:

> More broadly, in my mind, the concept of a global creative network, a cybernetic art matrix, was clear but not until some fifteen years after I had first digested the significance of integrative systems did I come upon the technology which could effect these transformations of culture I had so eagerly anticipated.
>
> (Ascott 1984: 16)

While Ascott was envisioning the creative possibilities of this new age, others like Paul Virilio were concerning themselves with the potential dark side of this phenomenon, analogous to the latent potential that high-speed trains have for resulting in high-speed rail disasters. Virilio wrote that to invent the sailing ship or the steamer is to invent the shipwreck, and to invent the family automobile is to produce the pile-up on the highway. As such, Virilio was concerned with the potential catastrophes that were to come from the onset of the global network age (2011: 10).

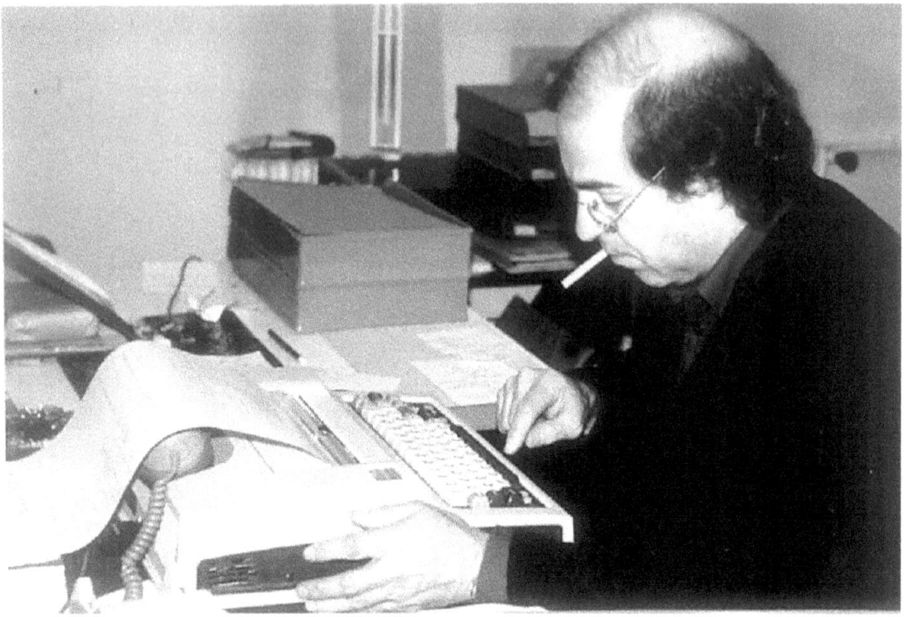

Figure 47: Roy Ascott working on the ARTEX network project, La Plissure Du Texte, 1980–83. Courtesy of the artist.

The network effect, the outcome of the dawning age of global networking, not only facilitated the sharing of consciousness but also, latently – like the derailed train – the inception of the digital or data virus. Where Virilio saw room for concern, Ascott's global network was one in which conceptual and physical exchanges of minds would fast-track the enrichment of culture and facilitate a globalist humanity. In Ascott's time of the 1980s, cyberspace was a phenomenon that was formative and still innocent. Ascott writes:

> Numerous writers have attempted to describe the enormous changes they see occurring in human awareness, which some see as a kind of planetary consciousness. Teilhard de Chardin imagined a noosphere, a thinking layer, enveloping the biosphere of the earth. Peter Russell has more recently advanced the hypothesis of the emergence of a planetary brain which may put us onto 'The threshold of a completely new level of evolution, as different from consciousness as consciousness is from life and life is from matter [...] A global brain which will result in a shift in human ego-centred awareness to a unified field of shared awareness'.
>
> (Ascott 1984: 50–51)

The global network age thus was perceived as a double-edged sword, with the power for global-sociocultural enrichment and cohesiveness on the one side, and destructive viral capabilities on the other. One way or the other in this new age, consciousness could be transported via data networks towards the instantaneous alteration of our native comprehension, of what we know about being in the world, with a global-scale reach that could theoretically perforate even the most isolated cities in the world.

In 1983, Roy Ascott was instigating a project entitled *La Plissure Du Texte*. There were a series of nodes across the world, run by Robert Adrian X (Vienna and Vancouver), Bruce Breland (Pittsburgh), Eric Gidney (Sydney), Norman White (Toronto), Helmut J. Mark (Vienna), Gregory McKenna and Tom Klinkowstein (San Francisco), David Garcia and Annie Wright (Amsterdam), and John Southworth (Honolulu), related to the ARTEX network. In a letter sent to the Sydney node and addressed to those colleagues who might help facilitate his envisioned project, he laid out the plan as follows:

> [*La Plissure Du Texte* is] a computer networking art project designed by Roy Ascott for the major exhibition devoted to the historical and contemporary review of Electricity in Art organised by Frank Popper for the Musee d'Art Moderne de la Ville de Paris. The exhibition is called 'ELECTRA 1983' and will open at the beginning of December I983 and run for three months. La Plissure du Texte is a collaborative story telling project using a computer timesharing network of artists located in Europe, North America and Australia. Artists using terminals in their own studios or at publicly accessible locations will be involved in a process of 'distributed authorship'. They will employ the ARTBOX network of I.P. Sharp's APL system (who have donated free network time to the project). The video display and print-out of this collaborative project in the form of a 'Planetary fairy tale' will constitute the contribution to ELECTRA.[14]

The global art project was already being visualized by Ascott and his peers, served as a collective challenge to the metaphorical atmosphere around to act as the host for 'connectivisim', in their words.[15]

Ascott prophesized in his paper 'Love in the Telematic Embrace' (1990) on the holistic nature of networks that new concepts of matter being explored would change people's perceptions:

> A telematics network is more than the sum of its parts, more than a computer communications web. The new order of perception it constitutes can be called 'global vision,' since its distributed sensorium and distributed intelligence – networked across the whole planet as well as reaching remotely into galactic space and deep into quantum levels of matter – together provide a holistic, integrated viewing of structures, systems, and events that is global in its scope.
>
> (Ascott 2003: 237)

In *A Computer in the Art Room* (1964), Catherine Mason writes about Ascott's beginnings in terms of his exploration of system thinking in art in the construction of works where the viewer becomes a co-author. Painting (expanded) in those early days was Ascott's chosen medium wherein marks were made on moveable pieces of glass whose latent interactivity made the viewer complicit in the creation of the artwork. The fact that Ascott used painting as the means through which he articulated the effects of emergent technologies on society is significant in defining the contemporary role of the artist.[16]

The subsequent remodelling of education programs for artists was predicated on the question of whether new technologies could serve as new media for artistic work, forcing the interrogation of new media in terms of the question: how will new technology change artistic practice?

The shift away from the face-to-face programs involving the master and the apprentice to a distributed network of global knowledge whereby the user could access information from any node in the network was seen by Ascott and his peers as primarily a cultural, and also a social and spiritual, outcome. The fact that Ascott used the lens of painting as a historical, theoretical and aesthetic cornerstone to his dealings with new technologies in the emerging age of global networks demonstrates the ongoing syncretic relationships between traditional practices of art and emerging technologies.

Ascott's visions for the future have been remarkably profound in creating aspirational perspectives, facilitating new ways of understanding the world and the human condition that incorporate the centrality of technological innovation. These futuristic visions are generous, freely given gifts, endemic of the time and at the cutting edge of syncretic contemporary art practices. The spread of ideas via the virtual networks has dominated the discourse of media and network theory as they manifest in the arts, facilitating new ways of questioning space, time and scale, and concepts of connectivity. It is in the engagement with emerging technologies that creative practitioners like Ascott have sought to reconceptualize and revitalize the role of the arts in informing what it means to not only be in the world (ontology) but the role technology plays, at the fundamental level, in shaping the human condition and its ecology.

The role that the arts can play in illuminating 'being' in the age of the global network and as such articulating the new culture of the data cloud, wherein these emerging technologies are exponentially redefining the social and economic framework of human civilization, is one sorely needed. Ascott pointed out in regards to the potential borne of research into telematics systems, for instance:

> In the context of telematic systems and the issue of content and meaning, the parallel shift in art of the status of 'observer' to that of 'participator' is demonstrated clearly if in accounts of the quantum principle we substitute 'data' for 'quanta.' Indeed, finding such analogies between art and physics is more than just a pleasant game; the web of connections between new models of theory and practice in the arts and the sciences, over a wide domain, is so pervasive as to suggest a paradigm shift in our world view, a redescription of reality and a recontextualization of ourselves. We begin to understand that chance and change, chaos and indeterminacy, transcendence and transformation,

> the immaterial and the numinous are terms at the center of our self-understanding and our new visions of reality.
>
> (Ascott 1990; 242)

In this reference from Ascott there is a synchronicity towards the significance of 'chance and change, chaos and indeterminacy, transcendence and transformation, the immaterial and the numinous' as being simultaneously contributing to a new understanding between art and science the new language of art. Ascott also makes a serious critical warning for art not to become lost in the euphoria of embracing emerging science and technology when he states:

> [T]here is a danger that the outcomes, even when beneficial in engineering, medical and social terms, could be spiritually hollow, and as such would exacerbate rather than relieve the excessive materialism of our time.
>
> (Ascott 2006: 1)

The contemporary data cloud is a strange, unruly metaphor, with its potential for folding and twisting, swelling and generating cultural turbulence, facilitating a paradigm shift in culture. The elemental cloud that is now the visual manifestation of data is seen as being a floating embodiment of turmoil that moves without the restrictions of perspectival space, which as such flows freely out in all directions. With no oppressive fields of containment, the cloud manifests like the unpredictable vagaries of the elements that, as a sum of an impossible collaboration of minute forces, form the upper atmosphere. The formal limits of clouds are unstable and ever changing; thus, they are unfathomable in their essence as a flux of unpredictability, a quality embraced by the Fluxus movement. The inherent chaos at the core of the data cloud now is colonized and reconstituted within the cloud's new role. There, it redistributes power, not in a messy and unpredictable fashion, but within the hierarchy of a perspectivalized phylum. The cloud as form-cum-concept necessarily informs emerging interpretations of pre-existing data: data that is scrambled and disarrayed by the unpredictability of movement of the cloud, which renders it as a semi-autonomous (non-) sorting algorithm, a perfect random generator. In demonstrating the need for artists to be involved with radical rethinking of creative strategies in confrontation with new technologies, Ascott writes:

> Our quality of life depends in large measure on finding creative and constructive value in the new technologies and in shaping them to human needs. Advanced telecommunications, digital multimedia, computer-mediated and post-biological systems are playing an increasingly important role in the transformation and globalisation of culture. Artists working in the space between art, science and technology are uniquely prepared to influence, shape and exploit these changes, and to radically rethink artistic and cultural strategies.[17]

Today, post-biological systems are brought to the fore within the metaphorical cloud that serves to image the physical manifestations of both digital and analogue particles. The dual process of biosemiotics between the DNA code and analogue proteins parallels the atmospheric principles of molecules and our bodily-information exchange between the pre-material blueprints manifest in material DNA proteins.[18] Drawing breath on a cold day sees the lungs fill with air, wherein an invisible cloud coalesces, only visibly forming once it is exhaled into the cold morning air. The molecular cloud and the physical body are the parallel sites whereby particles combine in the act of inhaling and exhaling. This analogy serves to image the latent interrelations between intelligent and organic systems and their interpenetrations (i.e. the intelligence of organics; the organics of intelligence). This code of cloud-ness that is embedded in our bodies, as in physical clouds, does not fit in a Cartesian space and thus it wants to escape the quantifiable linearity imposed upon it so as to enable its own free movement through digital networks and data clouds.

Repositioning the cloud

The data cloud is a symbol that is immutable to being hacked as its very existence is based on the vagaries of currents and turbulence. The haphazardness of the digital atmospherics sees the metaphorical molecular data to be completely randomized. The data appears to be indecipherable, but this illusion is only one of the political ideologies of the repositioning of perspectival control. The hacking of the mainframe is not to reshape the world but the pursuit of the perspectival goal, an unassailable hierarchical point of converging lines that in reality never converge and therefore cannot be reached.

To demonstrate the idea of the cloud and its conceptual framework as a way of making sense of the emergence of the digital-network paradigm, we can turn to those prescient art-historical moments and in particular to Turner and his watercolour cloud series (1820–30). Turner's paintings demonstrate a coalescence of materials, atmosphere and thought that, in hindsight, aided in the facilitation of a new molecular aesthetics. Turner's own clouds in many ways pre-empted the ideology that would inform, and in turn be produced by, the contemporary data cloud, through its use as a symbol of subversion and resistance to perspectival order and its parallel ideologies. Turner's clouds in effect anticipated (hacking) a change in the way power and authority is understood by manifesting an attack on the oppressive interests of perspectivalism.

The aesthetic molecular data cloud, as it rises to clear the landmass, releases its bits of data as rain from the heavens with their potential for new 'becomings'. The writings of Lucretius, in a poem entitled 'Of the Nature of Things', sought to articulate uncertainty, randomness and chaos, qua the atomic swerve, as the anti-rational heart informing all classical matter. Lucretius's atoms in the void, like raindrops that fall through the sum of the smallest perturbances in the atmosphere, re-enact the formation of new matter and meaning.

> Thuswise to swerve, down would they fall, each one,
> Like drops of rain, through the unbottomed void;
> And then collisions ne'er could be nor blows.
>
> (Carus 1950)

Once we perceive that which was at first invisible and thus inconceivable, and once those forces that inform our material world become visible and conceivable, our model of the material world shifts and shifts us in turn. We perceive that thinking and thoughts are atomic and stem from an intuitive understanding and rationalization of nothingness. In other words, if nothing cannot exist and everything is something and the smallest individual component of our physical universe is atomic, so too must thoughts be atomic in nature; our thoughts are made of the same fundamental material that those thoughts concern themselves with – the world at large.

When we consider the resonance between the Lucretius atomist conception of the world-making power of atomic swerve, and the uncertainty at the heart of quantum theory, we may also locate the emergent power of quantum computing as the most recent development in the genealogical development that links chaos to cognition. Thinking the world into existence becomes part of our anthropocentric attempts to understand the universe and our role within it. The thinking machines generated by the same atoms as their creators demonstrate what we already intuitively understand. From the biological to cellular to atomic-quantum, the same vibration and resonance is happening at the core of our physical universe. This embodied phenomenon cannot be separated into a fragmentation of individuals, but must be seen as fragments of the whole.

What is at stake with the emergence of atmospherics and molecular aesthetics is the officiation of an ultimate symbolic return to vagaries, ephemerality and immaterial existence beyond the rational controls of the perspectival order. A reinvention of our own intuition as a physical attribute and quality is now embodied in the 'electron cloud'.[19] This ephemeral and radical aesthetics is one we see expressed in the uncontrollable matter that populates the light-play in Turner's late works. There, in the burgeoning Industrial Revolution of Western Europe, we can perceive the future of matter as swarming molecular energies: the creation of atmosphere and the cloud. The movement of the atmosphere is not depicted as a static form whereby those elements that constitute it are discrete, but as formal mass inherently separate, beyond the controls of the pictorial-perspectival model of representation that defined image making in Turner's time. A cacophony of forces made visible in Turner's images predicts the looming end of the perspectival order; his clouds resonate with the language of atmospherics and the new direction of consciousness catalysed by the dawn of the atomic and the quantum.

In *An Introduction to Metaphysics*, Henri Bergson expands on the idea of the end of the sovereignty of the perspectival order, via his articulation of the notion and problem of infinite indivisibility:

> We believe that we can form a faithful representation of duration by setting in line the concepts of unity, multiplicity, continuity, finite or infinite divisibility, etc. There precisely

> is the illusion. There also is the danger. Just in so far as abstract ideas can render service to analysis, that is, to the scientific study of the object in its relations to other objects, so far are they incapable of replacing intuition, that is, the metaphysical investigation of what is essential and unique in the object.
>
> (Bergson 1912: 18)

What we see in Turner's painting is in essence an intuitive sense of the duration of those forces of physics that generate the world: a natural yet abstracted subject matter that demonstrates all the complexities of the world being fundamentally immeasurable by virtue of their exceeding complexity. Yet a sense of this complexity can be achieved via visual abstraction and symbolism. The cloud, in Turner's images and elsewhere, is a structure that is constantly both in flux and in transit – a structure that is by its very nature quantum, for it is susceptible to all the vagaries of measurement that defines the quantum condition.

What takes place when you observe an artwork that appears to naturally summarize what you actually see in the world? You ask the question: what would the world be like if this artwork was going to truly predict what you could see as being the world?

Turner's works painted just before his death demonstrate a culmination of years of observational knowledge mixed with the insights that were imparted to him through his friendships with prominent scientific thinkers of his time, such as Michael Faraday and Mary Somerville, bearing witness to landmark scientific experimentations and discoveries. In paintings such as *Rain, Steam and Speed – The Great Western Railway* (1844) (Fig. 48) and *Snow Storm – Steam-Boat off a Harbour's Mouth* (1842) (Fig. 49), Turner succeeded in distilling a prescient visualized sense of the emerging quantum-atomic paradigm that would define the decades to come.

Turner's Industrial Age, pictured in his canvases, is one characterized by the control of matter, and man's domination and exploitation of the material world and its resources. The unpredictability of the atmospherics and the clouds in Turner's works reflects the uncertainty of the future. *Rain, Steam and Speed* appears also to allude to the disintegration of perspective. The converging railway lines vanish in the coalescence of industrial and natural atmospherics (Fig. 48). The unassailable point is lost, the God-like vision becomes blurred – we only sense a future becoming conflated. These images are not glorifications of the Industrial Age but are a questioning of the residual effects that shift and change the landscape, signifying an 'un-becoming': a sensibility captured in the writing of Marshall Berman who stated, in regards to the condition of modernity, that 'all that is solid, melts into air' (1982: 13).

In Turner's time, scientific innovation was being translated directly into industry, which was transforming the physical landscape of the urban centres he moved between (from agrarian to industrial). This industrial landscape generated new demands for an appropriate visual language to communicate and critique the emerging industrial world as it became a misty, swirling, translucent atmospheric veil. Turner's courage was in forging an experimental visual language, expressed for example in *Rain, Steam and Speed* – the solid entity at the centre of the painting, the steam train, dissolves as it bellows out its smoke and

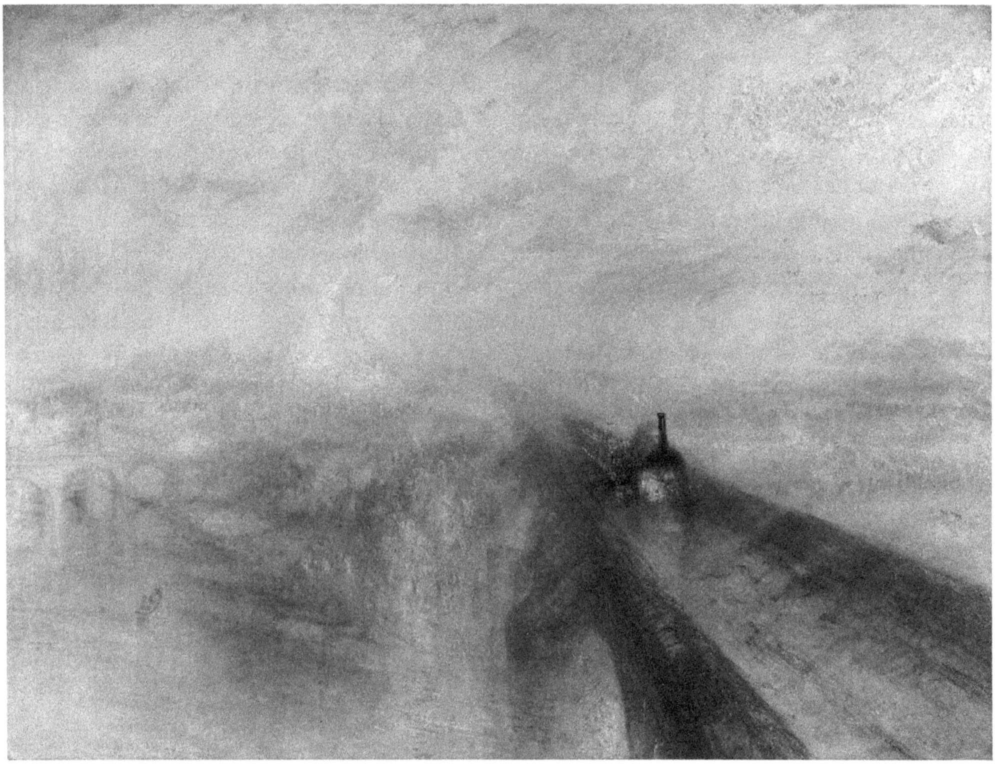

Figure 48: J. M. W. Turner, *Rain, Steam and Speed – The Great Western Railway*, 1844. Oil on canvas. Courtesy of National Gallery, London.

steam that simultaneously merges with the clouds, rain and the atmosphere. The dissolving mass and flux denying the vanishing point is the pertinent image of the age of uncertainty.

The oil paint, watercolour and drawn marks made by Turner show the cloud as a manifestation of a powerful force. In 1831, Michael Faraday demonstrated how placing iron filings on a piece of paper with a magnet underneath could reveal electromagnetic patterns, as if literalizing a line between electromagnetic conduit and conscious intuition. James Hamilton suggests that Turner's paintings were a direct representation of forces like these (1998: 126). These collective phenomena, which were developing from a range of scientific developments, gave Turner the sense that the world was not so much forming as a set of quantifiable and predictable patterns but rather by way of turbulence and uncertainty. This uncertainty manifests in the atmosphere of Turner's paintings, which do not allow for any such singular reading to take place.

The abstractions of energies in Turner's paintings are thus diagrammatical (Fig. 49) in the sense discussed in Chapter 2, as his paint seems to be feeling out the terrain of those forces underwriting the industrialization of the world. The painter's brush marks are intuiting the impending realization of uncertainty at the heart of the physical world – drawing from, among other developments, the complex patterns produced by magnetic filings once struck with a current.

Turner's fascination with magnetic filings and their radiating forms (dancing north and south along the lines of electromagnetic currents) effectively rendered them diagrammatic in their significance for him. It was this diagrammatic foundation that Turner used to build his works upon, as Francis Bacon did subsequently, to serve as conduit and analogy for something coming into existence. The diagram thus evolves through mark-making processes into a machinic concoction of everyday life. The magnetic pattern is not a single entity on a piece of paper, but rather part of a larger diagrammatic framework, drawn upon here for use in painting, involving the visualization of the normally invisible molecularization of all matter. The diagram plays an essential role, enabling the basis for intuition to take place. It does not force, prod or instruct intuition into occurring, but it does lay down the groundwork to enable a visible expression to take form. Intuition in this respect can be seen not as a single mathematical point in a diagram, but more like a writhing entanglement where any given point instantly informs its other partners.

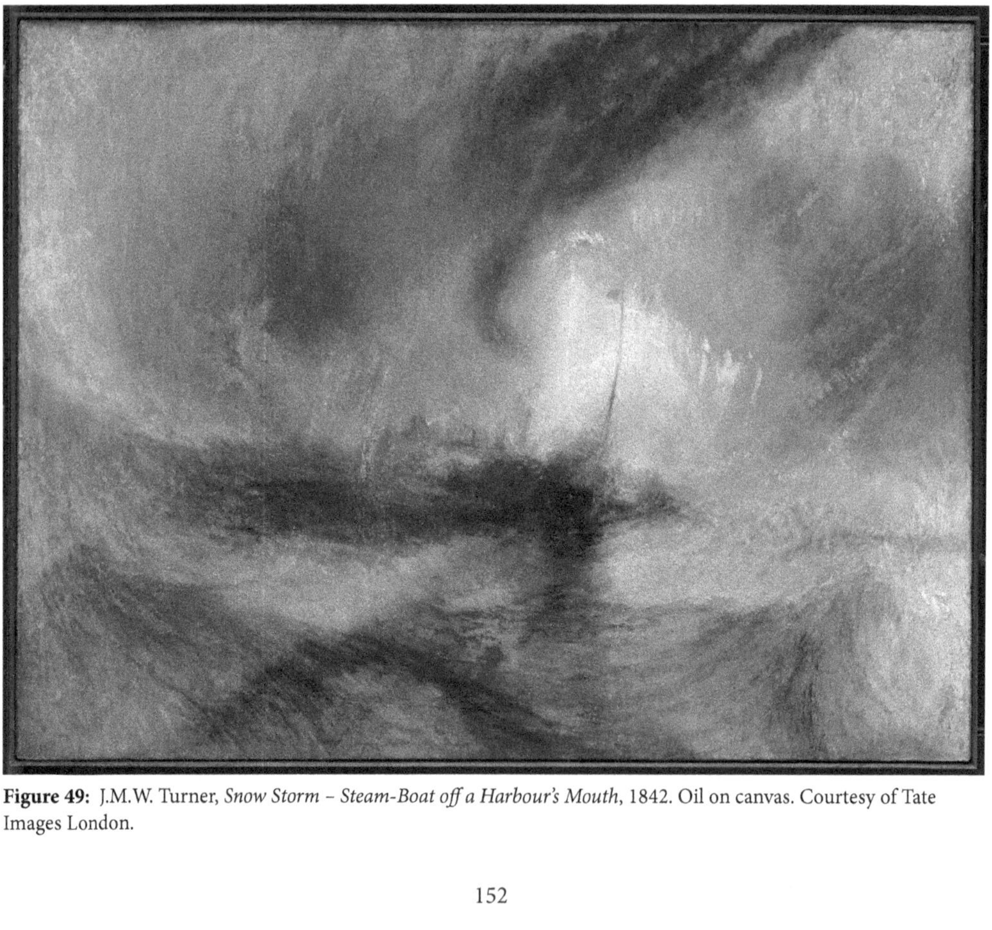

Figure 49: J.M.W. Turner, *Snow Storm – Steam-Boat off a Harbour's Mouth*, 1842. Oil on canvas. Courtesy of Tate Images London.

The intuitive mark, unlike the self-conscious mark, must happen where its effect and authority can be felt instantly, at 'infinite speed', throughout the artwork. Here we encounter the diagram as a conduit of forces – diagram as perspectivalist schema.

The schema of atmosphere, on the other hand, concerns that part of the invisible space that surrounds us. Atmospherics have a long history in physics, for example, the ether, which is the space that Maxwell suggested allows electromagnetic waves to vibrate in and across. This invisible and universal inter-medium formulated by Maxwell is, in his reckoning, the means by which all things are transmitted and received; the necessary conductive fundament through which all things are made manifest as physical information. The senses are constructed as a toolset for interpreting all of this information in some kind of practical way, for the survival of self-rendering the unknown into the known.

The spatial trends in painting and the emergent theories that were defining physics meant that at the turn of the twentieth century, art and science enjoyed a close relationship with the modern concept of ether. Ester Coen notes the prescience of painters by citing a passage from Umberto Boccioni's writings in his notes for a lecture given in 1911, in which Boccioni states '[w]hat needs to be painted is not the visible but what has heretofore been held to be invisible, that is what the clairvoyant painter sees' (1988: 239).

Boccioni reveals more about the creative nexus of the visible/invisible in the 1911 lecture where he posits that the condition of 'uncertainty weighs on the conscience of the modern artist' (Coen 1988: 239). This critical insight reveals a state of mind that serious artists were consumed with at the turn of century. The epistemological crisis brought about through the affirmation of Heisenberg's uncertainty in 1927 must therefore be understood as a gradual conjoining rather than a singularity.

The genealogy of a mechanistic intervention with materiality and physicality in the impending conditions of the First World War was witnessed three years after Boccioni's lecture. The devastating impact of machines on human and non-human manifestation was palpable. The impending comprehension of an unknowable quantum world was being enabled by physicists. Concurrently, art was the comprehension of a psychoanalytic conception of the world that was being inverted to a point of not being able to locate where the self resides. These three elements – war, quantum and psychology – in context, empowered the rise of uncertainty from what was a deterministic culture. Boccioni's acknowledgement of the weight of uncertainty meant that nature was to be irreversibly affected by the establishment of quantum uncertainty.

The machinic whole

If we could be better positioned to comprehend the atmospherics of the cloud and 'big data' through a molecular knowledge of the world, then this is another reason to change our engagement with materiality. We must first conceive of how we might instigate a change, conversation or transformation.

The new work by artistic collaboration Semiconductor,[20] titled *Earthworks* (2016) (Fig. 50), is a five-channel computer-generated animation. Semiconductor is the nom de plume for Ruth Jarman and Joe Gerhardt's collaborative practice. SonarPLANTA, who plan to develop new media and other emerging technological forms of practice in the cultural space, commissioned Semiconductor to create the artwork.

The impressive looking work is in an immersive landscape format. In its scale, the work tackles the enormity and complexity of nature. *Earthworks* attempts to visualize nature germinated from technical seismic data from sources such as glacier, earthquake and volcano, representing a new understanding of the global impact of change and uncertainty. Like Turner's awareness of the rising clouds of industrialization that was materializing new forms of environmental data, there is an ongoing attempt by the artists to capture new realities.

The sonification of the data works to create a sympathetic layer to the visualization, enabling the quantity of material being processed to dissolve into a transparent seminal layer. Semiconductor have developed a number of works that deal with scientific data and quantum phenomena. *Earthworks* reveals a continuation of their work dealing with the molecularization of the landscape.

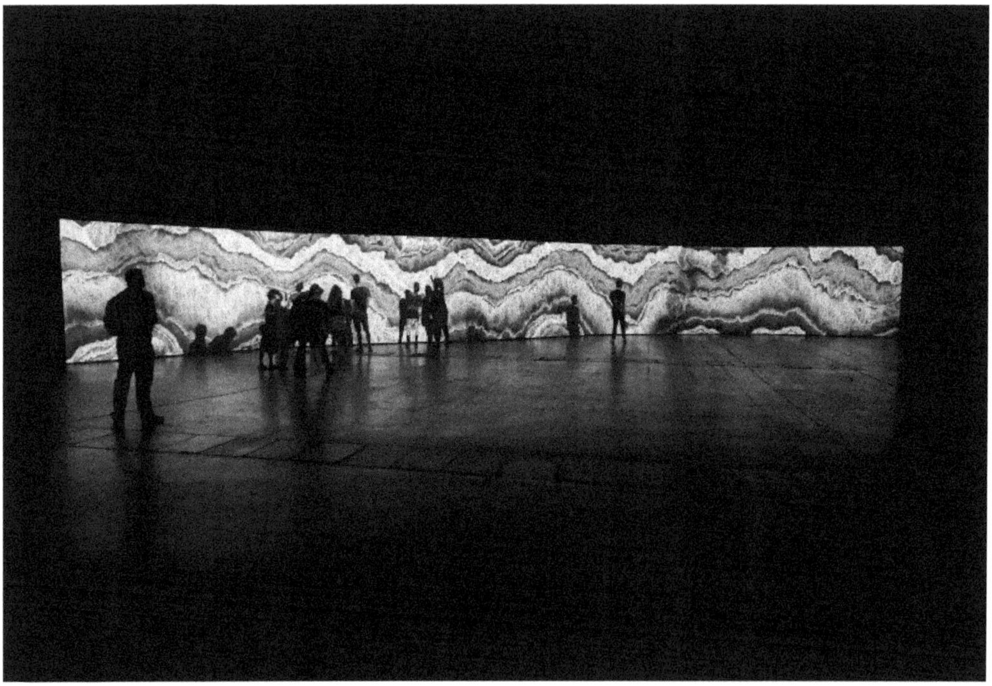

Figure 50: Semiconductor, *Earthworks*, 2016. Earthworks is commissioned by SónarPLANTA and produced by Advanced Music. Courtesy of the artists.

This landscape is where the world seems to transmute into its molecular states where chaos and uncertainty of reality are expressed by a sympathetic intuitive response that does not merely illustrate the phenomenon. Rather, the work attempts to create art that demands of the artists that they trust in a means of expressing information that in itself cannot be seen. The data turns into an atmospheric milieu that draws the observer along to see through the landscape and witness the concealed movements of the world taking shape (Fig. 50).

Engineer's perspective

The artists Oron Catts and Ionat Zurr who formed the group Tissue Culture and Art[21] make the case that we are caught seeing things through an instrumentalized perspective, as if through the eyes of an engineer. The artists write, '[o]ne way to emphasise and attract attention to alternative frames of thought is to open up the very same tools and spaces that serve this future to other disciplines, including artistic' (2012: 260).

The expanding scope of possibilities created by the instrumentalized present opens up the possibility of predicting the future, to question the very nature of causal effects borne out of perspectivalism and the way that we transform the world through a machinic logic. The premises and promises of an empirical knowledge led to a human-scale philosophical comprehension whereby the option of chance plays a major part in the way humans interacted with the material world. This machinic interaction is where the body-as-instrument is seen as being assembled from a few heterogeneous, but common, chemical compounds. These compounds are found in all other material objects that coexist in the planet. The potential for then seeing the world as being made of a synthesis of universal materiality goes a long way towards substantiating a vitalist theory of matter and materialism, a theory that has been articulated by Michel Serres, who writes:

> Greek wisdom reaches one of its most important points here, where man is in the world and of the world, in matter and of matter. He is not a stranger in the world but a friend, at home in the world, a fellow voyager, an equal.
>
> (Serres 1982: 121)

What Serres provides for us in his writing is a plan for the future, wherein all are equal: all mutually appreciative and self-identifying as universal travellers moving together, although each at a different speed. No one thing is more important than the other in this universalist and vitalist theory of the atmospherics of consciousness. In this model, when creative practitioners make images they do so to describe a vision of the world where the potential is for intuitively making connections with the phenomenal processes, laws that long predate consciousness, which we call chemistry physics. Without knowing it to be explicitly the case, all artists seek a total-world borne of a heightened intuition and comprehension of consciousness-as-nature as a coalescence of resonant, vitalist materialism. In so doing, creative practitioners manifest the latent power of consciousness as a force embedded in the universe it attempts to understand (not other to it, as premised in perspectivalism).

The world-model of anti-perspectival atmospheric uncertainty is bound up with the physical laws of quantum physics and the uncertainty of the world that it effectively distils. The materials that constitute the entity we know as 'human' are part of a shared universal resource over which the actual condition of humanness exerts no authority or monopoly. But because uncertainty, we now understand, is an (un)quantifiable fact of the physical universe, it is this very materiality of uncertainty that is latent in the human condition also. This uncertainty is a variable that is little acknowledged; how does the latently uncertain phenomenon of the physical world affect our awareness of being in it? This question is one that necessarily concerns both creative practitioners and scientific inquirers. I repeat the final line of Feynman's inspired speculation from his 1982 paper on what was at stake in getting to the quantum heart of things:

> And I'm not happy with all the analyses that go with just the classical theory, because nature isn't classical, dammit, and if you want to make a simulation of nature, you'd better make it quantum mechanical, and by golly it's a wonderful problem, because it doesn't look so easy.
>
> (Feynman 1982)

This summary of a very different quality of thinking about matter in the quantum age is part of the strange yet intuitive insight that inherently needs to be cultivated. This is the premise of the quantum leap between knowing and not knowing, wherein the post-perspectival world comes into its own and the pre-perspectival finds curious resonance. In quantum theory, the superposition presents itself as a sum of all possible outcomes of a given system. To stake a claim concerning one position with certainty renders it inherently false. Atmospherics, as opposed to perspectivalist thought, embraces and accounts for the latent insecurity and instability of knowledge, and it is thus the atmospherics of consciousness that is needed in the twenty-first century.

Notes

1 The idea of an experimental theoretical methodology is to develop an area of play around the interpretation and appropriation of cloud-ness. The play is in identifying, expanding, reflecting and linking theories with practice that can act as agents for breaking down boundaries.

2 'Hyperobject' is a term developed by Timothy Morten to describe events that are so large that they deny the local and demand a global attitude, for example climate change (Morten 2013).

3 Wil McCarthy's novel *Bloom* can be seen as analogous to the clouds, where a swarm concept becomes a boundary-less body. The swarm is made up of self-replicating nanobots called 'Mycora' or technogenic life, which have become a total cloudlike entity (Thomas 2013: 64).

4 The human race has escaped the expanding nanobot bloom and people are now referred to as 'immunities'. The Mycora (nanobots) describes itself as occupying 'contiguous space' and as being 'unpacked'. In one of the final scenes, the sapient Mycora, eating through the

spaceship, communicates with the crew and states ‘there is a system not just bodies, no, course not. We do not have “bodies” in the sense you probably mean, but there are the complexes which constitute us, and the complexes which support us’ (McCarthy 1998: 303).

5 Deleuze and Guattari oppose the machinic, on the one hand, to the mechanical. The machinic applies to the machine as a functional unity of discrete but homogeneous parts, and on the other to the organic, which applies to the organism as a hierarchical organization of biological organs. The assemblage itself is not opposed to either mechanical machines or organic bodies but encompasses both. Bodies and machines enter into machinic relationships, that is, become parts of an assemblage (Johnston 1999).

6 In Manetti’s biography of Brunelleschi, the account of the perspective panel (peephole device), dated around 1413, features in a small but highly significant way. Although written some years after Brunelleschi’s death, the reliability of Manetti’s insights is reflected in the fact that he visited the home of Brunelleschi when he was alive and was also shown the peephole device. Along with Brunelleschi, Manetti was interested in geometry and mathematical sciences and could have picked up a lot of his information about the device directly from their discussions. Suggesting that Brunelleschi was proud of his perspective panels and glad to show them off, Manetti stated, ‘I have had it in my hands and seen it many times in my days and can testify to it’. The device had three components. The first of these, as Manetti describes, comprised a ‘small panel about half a braccia square’, upon which was a painting of the Baptistery of San Giovanni (adjacent to the front of the cathedral of Florence), ‘painted with such care and delicacy and with such great precision’. The second part of the device was a ‘flat mirror’, which was to be held in the other hand to that of the painting. The third part of this device was the small hole cut into the panel itself: ‘he had made a hole in the painted panel at that point in the temple of San Giovanni which is directly opposite the eye of anyone stationed in the central portal of Santa Maria del Fiore’ (Thomas 2006).

7 In *Renaissance and Renascences in Western Art*, Panofsky uses the terms ‘quantum continuum’ and ‘quantum discretum’ to define the different uses of perspectival space when dealing with continuous or discrete spatial oriented objects. The vanishing point is a single entity that is discrete but it is the effect of the continuum. It is the illusion of control where the point is fundamentally indivisible from, and therefore part of, the continuum (Panofsky 1944: 126).

8 The use of the other in this document fits well into the cloud concept as an ontological problem between being and not being. A liminal space that enables two states to exist at the same time without denying the other.

9 From Philip K Dick’s novel VALIS (acronym for Vast Active Living Intelligence System), ‘a perturbation in the reality field in which a spontaneous self-monitoring negentropic vortex is formed’ explains a situation where the reverse of entropy takes place and things become more ordered (Dick 1981).

10 Entanglement is used in physics to mean where a particle cannot be seen as being independent of another particle.

11 This project was happening at the same time as Alain Aspect published his paper on Bell’s inequality, which suggests measurements made by observers affect pairs of particles that are entangled and separated. Like *Hole in Space,* it explores local realism by creating challenging assumptions of relationships between people that become part of the phenomena of being entangled.

12 Paul Sermon was an ex-student of Roy Ascott, studying from 1985 to 1988 for a B.A. Honours Fine Art Degree at the Newport School of Fine Art at Gwent College of Higher Education. Ascott was developing a new concept for art education, involving the integration of new media, when Sermon became interested and involved in telematic artwork.

13 In his book, *Information Arts: Intersections of Art, Science, and Technology*, Simon Wilson states that 'Telematic Dreaming allows not necessarily an escape from the body but the opportunity to observe oneself from a new perspective' (Wilson 2001: 520).

14 This is a section of a letter from Roy Ascott to Eric Gidney in Sydney as part of the international networked participants of the art project *La Plissure Du Texte*.

15 'Connectivism' is a term created by Ascott to suggest that an artist working and making in the new area of art and telecommunications was actually an art movement (Ascott 2013: 308).

16 The connections are clearly articulated by Ascott when he suggested that the 'most powerful metaphor of telematic interconnectedness that painting of that transitional period could offer is to be found in Jackson Pollock's work [...] He creates "fields" of intertwining, interweaving, branching, joining, crossing, linking lines of energy, creating an alloverness and allatonceness which is the very epitome of telematic networking' (Ascott 1984: 47).

17 This extract is from Ascott's Centre for Advanced Inquiry in the Interactive Arts (Caiia) course documentation from 1994.

18 Jasper Hoffmeyer and Claus Emmeche suggest that biosemiotics 'is the basis of this self-reference, and thus the basis of life? We shall suggest here that the central feature of living systems allowing for self-reference, and thus the ability to select and respond to differences in their surroundings, is code-duality i.e. the ability of a system to represent itself in two different codes, one digital and one analog' (Hoffmeyer and Emmeche 1991).

19 The electron when understood metaphorically as a cloud is a way to visualize a group of electrons moving around the nucleus of an atom. The cloudlike cluster of electrons fits well with the uncertainty that exists at the quantum level. Schrödinger (1926), reputedly building from Bohr's initial diagram of the atom, was one of the first to come up with a cloudlike visual metaphor to deal with the inability to measure the actual position of an electron.

20 The Semiconductor website states that Jarman and Gerhardt 'make visually and intellectually engaging artworks which explore the material nature of our world and how we experience it through the lens of science and technology, questioning how these devices mediate our experiences' (Jarman and Gerhardt 2017).

21 Tissue Culture and Art was the biological art group that founded SymbioticA. SymbioticA, situated in the University of Western Australia, was the first arts lab attached to a human biology department that allows artistic residencies to work with tissue culture as a material and medium for making art. The creation of a semi-living material shifted the perspective of art to align with the emerging philosophical movement of New Materialism. Their work has been exhibited at Ars Electronica in 2001 in the exhibition *Takeover*, and at a retrospective of their three main projects at the Biennale of Electronic Arts, Perth, in the exhibition titled *Biofeel* in 2002.

Conclusion

Fragments

I conclude this book by elaborating on the phenomenon of fragmentation and the fragments of thought that constitute the makeup of the previous chapters; how do these atomistic fragments of text permit us glimpses into alternative ways to rethink, reshape desire, reorient hope and reimagine possibility of being in the world? These fragments of ideas in the text appear to disperse and reform through their own indeterminacy, resonating with the quantum condition these very words have sought to reveal. It seems as if consciousness operates through a shared medium with the physical plane such that thought is in essence delivered through waves, magnetism and gravity, which, in turn, are translated by humans into an experience of something external. Therefore, all fragments – emanations of quantum atmospherics – are only existing through a process of scale and proximity. Take a broken cup as it lies on the floor – Susskind, in his lecture 'Why is Time a One-Way Street', suggests that the laws of physics permit it to be put back together from the fragments that were dispersed when broken, in effect reversing the pull of entropy. There is, Susskind proffers, a latent probability that the broken pieces could put themselves back together if all the molecules in the shattered cup, the floor and atmosphere around the cup were to move in an opportune fashion. Here, I propose an image of the intellectual labourer moving both through and against the invisible tides of probability: an image of the product of intellectual labour, such as this book, as a cup fluctuating between probable states of order.

Paalen once put forward a dramatic idea of creative production, paraphrased by Parkinson on the matter:

> Insists that the artist's sense-led, intuitive, qualitative comprehension of nature should be accepted alongside the scientist's 'objective,' abstract, quantitative investigation of the same, in an overarching theory of knowledge.
>
> (Parkinson 2004: 576)

Yet in the age of the quantum – the age of uncertainty – the imposed gap between creative and scientific inquiry is diminishing as the role of intuition becomes the necessity of thought to navigate where and what it cannot know with certainty. Could we comprehend differently quantum phenomena from outside the entrenched ocular-centrism of consciousness? This book has sought to embrace an Epicurean and pre-perspectival view of the universe to make sense of this dawning post-perspectival condition.

Artists have historically embedded themselves in, or moved in and out of, different disciplines with the hope of giving image to the world. As they encounter new insights, the results of artists are as such often estranged; their findings cannot be readily understood because, when probing the limits of knowledge, the necessary breaks in language limit the readiness of comprehension. Avant-garde artists seek to undermine our inherited and preconditioned world-aesthetics. The artists discussed in this book know they need to disturb the order of knowledge in order to articulate the new. Meanwhile, contemporary physicists still talk about not understanding quantum mechanics developed in 1927, even though it has been the most robust and successful theory ever developed. Quantum mechanics explains the world as being based on probability and chance, empowering what cannot be defined in singular deterministic models. Physicists still look for the single equation that could summarize the whole of the universe. Each physicist looks at the previous equations and adjusts to refine them, not to simplify the outcomes but to interconnect them and, if need be, break them.

Part of the historic function of artistic analysis has been the championing of fallibility. That is to say that avant-garde artists have unabashedly embraced uncertainty as the core of the human condition. The term human fallibility is not seen here in a negative context, but rather the inescapable character of consciousness. Thus, to embrace fallibility and uncertainty is to enable new languages to evolve and aid in our comprehension of the world. Our latent inability to represent the world as it appears to be, 'in which both object and subject seem less comprehensible', gives rise to an ever-evolving project of imaging produced ad infinitum by our fallible consciousness (Parkinson 2004: 560).

The eternal drive to develop new languages of expression is the imperative of creative inquiry. It is this imperative that led to the successive ruptures fuelled by the onset of modernity – such as Malevich's *Black Square* (1915). In their historical context, images like *Black Square* were meant to negate the limiting influence of deterministic, perspectival images of the world. It is the destabilizing effect of the avant-garde that has embraced the way for new knowledge to become recognized and therefore legitimized.

The culmination of the concepts for this book appear as a series of inceptions, like a living swarm of quantum self-actualizing mini-bots; all the thoughts, concepts, ideas and elements are in the mutual, atomic and collective process of becoming. Feynman tried to explain to the audience in his 1969 talks about quantum mechanics that the actual way nature works will appear to be so 'shocking' that he realizes nobody will understand what he is saying, not even himself (1979). Parkinson sought to qualify this 'shocking' reality in the production of new knowledge and the arts but could not yet see its resolution. He was particularly focused on the art and context of Paalen and Matta when he anticipated the paradigm shift that would eventually occur. Parkinson's quote is reintroduced here to reiterate the need for more significant collaborations:

> The broad-ranging exchange that took place within the avant-garde and academia in 1930s France on the new challenges laid to traditional forms of rationalism by modern physics, psychology, philosophy, and art, is one of those awaiting a reception in the scholarship.
>
> (Parkinson 2004: 557)

In Bergson's writings, the conceptual approaches of philosophy are entwined with physics, dating back to 1889 in his book *Time and Free Will*. This was Bergson's first literary work; his philosophical concepts antedate by 40 years the intuitions of Niels Bohr and Werner Heisenberg on physical interpretation of waves (Broglie 1969: 46). Once more we see a historical convergence of creative intuition and scientific inquiry. Today, CERN's Hadron Collider, a monumental testament to the fundamental importance of particle physics, has revealed to us the physical existence of the long-theorized Higgs boson. First posited in the 1960s, the discovery of the Higgs boson indicates the ever-increasing capabilities of human technology to reveal alien elements of our physical world. Discoveries like the Higgs boson force a process of reconceptualization, as assumptions and securities we once held about the world are erased from beneath us.

The creation and development of 'quantum' artwork is the creative imperative of the twenty-first century. Only through creative and experimental dealings with the physical fact of quantum uncertainty will a paradigm shift in consciousness be brought to bear. The twenty-first century marks the explicit move towards an inculcation of a quantum consciousness. This book was written in the hope of contributing to the age of quantum consciousness. The strange and uncomfortable task of trying to develop a new way of comprehending and mapping the latently incomprehensible and un-mappable is the underlying task of this volume.

Quantum theory is forcing us to reconceive of the world as an alien landscape beset with an ontology of uncertainty that does not simply go far beyond the limits of our (twentieth-century) physics models, but goes against the grain of the rational logic that guides our understanding. We must look for alternative ways to calculate what is going on. Everything we see and experience through the paradigm of the classical world up to the revelations of the Enlightenment, we must now resituate as over-narrowed and in denial of fundamental expressions of quantum phenomenon.

Measurement features frequently in this book as a contestable form of understanding what constitutes physical reality. The beauty of a non-deterministic world is the open-ended search for meaning. If we were to discuss a thought experiment from Zeno's paradox,[1] we would see that it is only as strong as our understanding of the meaning and context of the concepts articulated. We would state that each scholar's interpretations that contradicted Zeno also, in turn, by default, contradict each other. The irony is that when a thought experiment captures the public imagination, the poignancy of its preponderance will be challenged. The initial thought experiment is actually problem finding and not problem solving, a provocation – its role is to stimulate more thoughts, to render alternative experiments. That is, it is the agency of uncertainty working as problem finding, as much as it is about acquiring solutions.

In Edgar Allen Poe's short story *A Descent into the Maelstrom*, two sailors and their boat are caught in a sudden onset of a ferocious storm and sucked down into a whirlpool. As every coherent object is rapidly destroyed and sucked down into the oceanic black hole, one sailor recognizes that certain artefacts from the boat keep reappearing. Only by attaching himself to one of these objects, by virtue of this ad hoc application of his knowledge, he saves himself from the whirlpool (2000). Marshall McLuhan uses this story to suggest that pattern recognition in the face of total chaos is the only way to survive. The story relates to an empirical, ongoing understanding of measuring and summing the world as it is – and indeed, as it could be. The shifting sands, like the whirlpool, all become analogous to perceiving patterns, rhythms and vibrations to attempt to bring the world into a classical form. The work of art takes shapes from all these factors that contribute to the strange love we have for understanding things beyond ourselves, which test the fundamental context of why we are here and what we are learning from the ontology of being.

This book has sought to mobilize knowledge to not only cope, but to thrive in the age of quantum – the age of uncertainty. Like Poe's sailor, I have attempted to facilitate, in an ad hoc, fragmented fashion, a new sense of the world through the ontology of uncertainty. Only by cultivating a consciousness borne of the quantum, the difficult and alien revelations of quantum theory, can we fundamentally rewrite and restructure a sense of our world and our place within it. In like fashion, only by accepting and working with the hazards and insecurity of the maelstrom was Poe's sailor able to return to some sense of stability.

The new aesthetic of the quantum age is one set against, and derived from, the existential challenge and anxiety that is posed to us by the epistemic uncertainty of quantum physics. If there is beauty in truth, then the beauty of the quantum age is hard won and insecure. The ontology of uncertainty is that which is at stake in the quantum age – from there is derived the aesthetics and the necessary consciousness of navigating quantum uncertainty. Only through the cultivating of quantum consciousness may we perceive the latent beauty of this ontological condition. This book has hoped to illuminate the quandary of quantum uncertainty. As a coalition of fragments, this volume of experimental ideas, histories and concepts is itself as much a product of, as a producer of, the age of uncertainty.

Note

1 Zeno's thought experiments the 'Arrow paradox' and 'Achilles and the tortoise' suggest that whilst measuring the flight of an arrow, or a running motion, they become a series of frozen instances in space. These instances are similar to 1+1+1 to be understood as 3, but we do not individuate or 'choose to make use of the multiplicity which is enclosed in each of the units' (Bergson 1910: 81).

References

Amato, I. (1999), *Nanotechnology: Shaping The World Atom by Atom*, Washington, DC: National Science and Technology Council and The Interagency Working Group on Nanoscience Engineering and Technology, http://www.wtec.org/loyola/nano/IWGN.Public.Brochure/. Accessed 11 January 2007.

Apollonio, U. (ed.) (1973), *Futurist Manifestos*, London: Thames & Hudson.

'art, n.1' (2017), *OED*, Oxford: Oxford University Press.

Ascott, R. (1984), 'Art and telematics: Towards a network consciousness', in H. Crundmann (ed.), *Art and Telecommunications*, Vancouver: The Western Front, pp. 25–67.

——— (1990), 'Is there love in the telematic embrace?', *Art Journal*, 49:3, pp. 241–47.

——— (2000), 'Edge-life: Technoetic structures amid moist media', in R. Ascott (ed.), *Art, Technology, Consciousness: Mind@large*, Bristol: Intellect.

——— (2003), *Telematic Embrace: Visionary Theories of Art, Technology, and Consciousness* (ed. E. A. Shanken), Berkeley and London: University of California Press.

——— (2006), 'Technoetic pathways toward the spiritual in art: A transdisciplinary perspective on connectedness, coherence and consciousness', *Leonardo*, 39:1, pp. 65–69.

——— (2013), 'Back to nature II: Art and technology in the 21st century', in M. More and N. Vita-Mote (eds), *The Transhumanist Reader*, New York: Wiley-Blackwell, pp. 438–49.

Aspect, A., Dalibard, J. and Roger, G. (1982), 'Experimental test of Bell's inequalities using time-varying analyzers', *Physical Review Letters*, 49:25, pp. 1804–07.

Barad, K. (1996), 'Meeting the universe halfway: Realism and social constructivism without contradiction', in L. H. Nelson and J. Nelson (eds), *Feminism, Science, and the Philosophy of Science*, vol. 256, Dordrecht: Kluwer Academic Publishers, pp. 161–94.

——— (2003), 'Posthumanist performativity: Toward an understanding of how matter comes to matter', *Signs*, 28:3, pp. 801–31.

——— (2007), *Meeting the Universe Halfway: Quantum Physics and the Entanglement of Matter and Meaning*, Durham: Duke University Press.

——— (2012a), 'Intra-actions: Interview with Adam Kleinman', *Mousse*, 34, pp. 76–81.

——— (2012b), 'On touching – The inhuman that therefore I am', *Differences: A Journal of Feminist Cultural Studies*, 25:3, pp. 206–23.

——— (2012c), 'What is the measure of nothingness? Infinity, virtuality, justice', in C. Christov-Bakargiev (ed.), *dOCUMENTA (13): The Book of Books*, Ostfildern: Hatje Cantz.

Bell, J. (1964), 'On the Einstein Podolsky Rosen Paradox', *Physics*, 1:3, pp. 195–200.

Benjamin, S. C., Ardavan, A., Briggs, G. A. D., Britz, D. A., Gunlycke, D., Jefferson, J., Jones, M. A. G., Leigh, D. F., Lovett, B. L., Khlobystov, A. N., Lyon, S. A., Morton, J. J. L., Porfyrakis, K., Sambrook, M. R. and Tyryshkin, A. M. (2006), 'Towards a fullerene-based quantum computer', *Journal of Physics: Condensed Matter*, 18:21, pp. S867–83.

Bennett, J. (2010), *Vibrant Matter: A Political Ecology of Things*, Durham: Duke University Press.

Bennett, M. J. (2013), 'Deleuze and Epicurean philosophy: Atomic speed and swerve speed', *Journal of French and Francophone Philosophy*, 21:2, pp. 131–57.

Bergson, H. (1910), *Time and Free Will: An Essay on the Immediate Data of Consciousness* (trans. F. L. Pogson), London: George Allen and Unwin.

——— (1911), *Creative Evolution* (trans. A. Mitchell), New York: H. Holt & Company.

——— (1912), *An Introduction to Metaphysics* (trans. T. E. Hulme), New York: Putnam's Sons.

——— (1949), *An Introduction to Metaphysics* (trans. T. E. Hulme), New York: Liberal Arts Press.

——— (1990), *Matter and Memory* (trans. N. M. Paul and W. S. Palmer), New York: Zone Books.

Berman, M. (1982), *All That Is Solid Melts into Air: The Experience of Modernity*, New York: Simon and Schuster.

Braun, M. (1992), *Picturing Time. The Work of Etienne-Jules Marey (1830–1904)*, Chicago and London: The University of Chicago Press.

Broglie, L. d. (1969), 'The concepts of contemporary physics and Bergson's ideas on time and motion', in P. A. Y. Gunter (ed.), *Bergson and the Evolution of Physics*, 1st ed. , Knoxville: University of Tennessee Press, pp. 45–62.

Burnham, J. (1968), *Beyond Modern Sculpture: The Effects of Science and Technology on the Sculpture of This Century*, New York and London: G. Braziller and Allen Lane The Penguin Press.

Callen, A. (2000), *The Art of Impressionism: Painting Technique & the Making of Modernity*, New Haven, CT: Yale University Press.

Carus, T. L. (1950), *Of the Nature of Things* (trans. W. E. Leonard), New York: Dutton.

Catts, O. and Zurr, I. (2012), 'Life as a raw material: Illusions of control', *Somatechnics*, 2:2, pp. 250–62.

Cézanne, P. (1976), *Paul Cézanne: Letters* (ed. J. Rewald), 4th rev. ed. and enl. ed., Oxford: Cassirer.

——— (2001), *Conversations with Cézanne*, Berkeley: University of California Press.

Coen, E. (1988), *Umberto Boccioni*, New York: Metropolitan Museum of Art.

Colin St John Wilson (1999), *The Artist at Work: On the Working Methods of William Coldstream and Michael Andrews*, Philadelphia: Philadelphia Museum of Art.

Collis, R. (ed.) (2012), *Graphite*, London: GV Art.

Corbett, J. M. (2008), 'Towards neuroscientific management? Geometric chronophotography and the thin-slicing of the labouring body', *Management & Organizational History*, 3:2, pp. 107–25.

Cotkin, G. (2015), *Feast of Excess: A Cultural History of the New Sensibility*, New York: Oxford University Press.

Dagognet, F. (1992), *Etienne-Jules Marey: A Passion for the Trace*, New York and Cambridge, MA: Zone Books and MIT Press.

Damisch, H. (2002), *A Theory of /Cloud/: Toward a History of Painting* (trans. J. Lloyd), Redwood City, CA: Stanford University Press.

Deleuze, G. (1993), *The Fold: Leibniz and the Baroque* (trans. T. Conley), London: The Athlone Press Ltd.

——— (1999), *Foucault* (trans. S. Hand), London: Continuum.

——— (2005), *Francis Bacon: The Logic of Sensation* (trans. D. W. Smith), Minneapolis, MN: University of Minnesota Press.

Deleuze, G. and Guattari, F. (1987), *A Thousand Plateaus: Capitalism and Schizophrenia* (trans. B. Massumi), Minneapolis, MN: University of Minnesota Press.

DeLys, S., Dyson, F., Hinterding, J. and Rose, J. (1996), 'Contributors' notes', *Leonardo Music Journal*, 6, pp. 111–15.

Dick, P. K. (1981), *Valis*, New York City: Bantam Books.

Drexler, K. E. (1986), *Engines of Creation*, 1st ed., Garden City: Anchor Press and Doubleday.

Einstein, A., Born, M. and Born, H. (1971), *The Born-Einstein Letters: Friendship, Politics, and Physics in Uncertain Times – Correspondence between Albert Einstein and Max and Hedwig Born* (trans. I. Born), new ed., Basingstoke: Macmillan.

Everett, H. (1957), '"Relative State" formulation of quantum mechanics', *Reviews of Modern Physics*, 29:3, pp. 454–62.

——— (1973), 'The theory of the universal wave function', in B. S. DeWitt (ed.), *The Many-worlds Interpretation of Quantum Mechanics*, Princeton: Princeton University Press, pp. 3–140.

Ference, M., Lemon, H. B. and Stephenson, R. J. (1956), 'Analytical experimental physics', *Physics Bulletin*, 7:11, pp. 623–24.

Feuer, L. S. (1974), *Einstein and the Generations of Science*, New York: Basic Books.

Feynman, R. (1956), 'The relation of science and religion: Some fresh observations on an old problem', *Engineering and Science*, 19:9, pp. 20–23.

——— (1960), 'There's plenty of room at the bottom', *Caltech Engineering and Science*, 23:5, pp. 22–36.

——— (1964), 'Probability and uncertainty: The quantum mechanical view of nature', *The Character of Physical Law*, http://www.cornell.edu/video/richard-feynman-messenger-lecture-6-probability-uncertainty-quantum-mechanical-view-nature. Accessed 11 October 2011.

——— (1979), 'Fits of reflection and transmission – Quantum behaviour', *The Douglas Robb Memorial Lectures*, Bristol: The Vega Science Trust.

——— (1982), 'Simulating physics with computers', *International Journal of Theoretical Physics*, 21:6–7, pp. 467–88.

Feynman, R. P. (1963), 'Quantum behavior', in R. B. Leighton and M. Sands (eds), *The Feynman Lectures on Physics*, vol. 3, Reading: Addison-Wesley.

Fisher, M. P. A. (2015), 'Quantum cognition: The possibility of processing with nuclear spins in the brain', *Annals of Physics*, 362, pp. 593–602.

——— (2017), 'Are we quantum computers, or merely clever robots?', *International Journal of Modern Physics B*, 31:7, n.pag.

Freud, S. (1927), *The Ego and the Id / by Sigm. Freud* (trans. Joan Riviere), London: Leonard & Virginia Woolf at the Hogarth Press, Institute of Psycho-analysis.

Frichot, H. l. N. (2011), 'Drawing, thinking, doing: From diagram work to the superfold', *ACCESS: Critical Perspectives on Communication, Cultural & Policy Studies*, 30:1, pp. 1–10.

Greene, B. (2014), 'Measure for measure: Quantum physics and reality', *World Science Festival*, https://www.youtube.com/watch?v=GdqC2bVLesQ. Accessed 26 February 2018.

Hamilton, J. (1998), *Turner and the Scientists*, London: Tate Gallery Publishing.

Henderson, L. D. (2004), 'Editor's introduction: I. Writing modern art and science – An overview; II. Cubism, futurism, and ether physics in the early twentieth century', *Science in Context*, 17:4, pp. 423–66.

——— (2013), *The Fourth Dimension and Non-Euclidean Geometry in Modern Art*, Cambridge: MIT Press.

Hinterding, J. (2012), 'Simple forces', http://www.haineshinterding.net/2012/12/30/simple-forces/. Accessed 26 February 2018.

——— (2016), interviewed by P. Thomas, [DATE].

Hobbs, C. A. (2008), 'Peirce's tychism and the epicurean swerve', *Southwest Philosophical Studies*, 30:Spring, n.pag.

Hoffmeyer, J. and Emmeche, C. (1991), 'Code-duality and the semiotics of nature', http://www.nbi.dk/~emmeche/coPubl/91.JHCE/codedual.html. Accessed 26 February 2018.

Jaeger, G. (2009), 'Double-slit experiment (or two-slit experiment)', in D. Greenberger, K. Hentschel and F. Weinert (eds), *Compendium of Quantum Physics*, Berlin and Heidelberg: Springer, pp. 174–78.

Jarman, R. and Gerhardt, J. (2017), 'Semiconductor', http://semiconductorfilms.com/. Accessed 26 February 2018.

Johnston, J. (1999), 'Machinic vision', *Critical Inquiry*, 26:1, pp. 27–48.

Kaiser, D. (2005a), *Drawing Theories Apart: The Dispersion of Feynman Diagrams in Postwar Physics*, Chicago: University of Chicago Press.

——— (2005b), 'Physics and Feynman's diagrams', *American Scientist*, 93:2, pp. 156–65.

Kandinsky, W. (1979), 'Point and line to plane', in H. Rebay (ed.), New York: Dover Publications.

Klee, P. (1961), *Paul Klee: The Thinking Eye (The Notebooks of Paul Klee)*, vol. 1 (ed. J. Spiller), London: Lund Humphries.

——— (1968), *Pedagogical Sketchbook*, London: Faber.

Konstan, D. (2014), 'Epicurus', *The Stanford Encyclopedia of Philosophy*, http://plato.stanford.edu/archives/sum2014/entries/epicurus/. Accessed 26 February 2018.

Kuh, K. (1962), *The Artist's Voice: Talks with Seventeen Artists*, New York: Harper & Row.

Lambert, N., Chen, Y., Cheng, Y., Li, C. M., Chen, G. and Nori, F. (2013), 'Quantum biology', *Nature Physics*, 9, pp. 10–18.

Laporte, P. M. (1988), 'Cubism and relativity with a letter of Albert Einstein', *Leonardo*, 21:3, pp. 313–15.

Laughton, B. (2009), 'Coldstream and Giacometti in London', *The British Art Journal*, 9:3, pp. 79–85.

Manetti, A. d. T. (1970), *The Life of Brunelleschi* (trans. C. Enggass), University Park: Pennsylvania State University Press.

Marey, É.-J. (1895), *Movement* (trans. E. Pritchard), New York: D.Appleton & Company.

Marks, J. (2010), 'Materialism', in A. Parr (ed.), *The Deleuze Dictionary*, rev. ed., Edinburgh: Edinburgh University Press, pp. 158–60.

Mattuck, R. D. (1976), *A Guide to Feynman Diagrams in the Many-Body Problem*, 2nd ed., New York: McGraw-Hill.

McCarthy, W. (1998), *Bloom*, London: Millennium.

McLuhan, M. (2003), *Understanding Media: The Extensions of Man* (ed. W. T. Gordon), Corte Madera: Gingko Press.

McLuhan, M., and Parker, H. (1968), *Through the Vanishing Point*, New York: Harper & Row.

Menezes, M. d. (2001), 'Visual Representation in Art and Science: A Study of Richard Feynman's and Joseph Beuys' Diagrams', MSt dissertation, Oxford: University of Oxford.

Merleau-Ponty, M. (1968), *The Visible and the Invisible* (trans. A. Lingis), Evanston: Northwestern University Press.

Miller, A. I. (1986), *Imagery in Scientific Thought: Creating 20th-century Physics*, Cambridge: MIT Press.

——— (1994), *Early Quantum Electrodynamics*, Cambridge: Cambridge University Press.

——— (2004), 'Einstein, Picasso', *Physics Education,* 39:6, p. 484.

Morello, A., Pla, J. J., Zwanenburg, F. A., Chan, K. W., Tan, K. Y., Huebl, H. and Dzurak, A. S. (2010), 'Single-shot readout of an electron spin in silicon', *Nature,* 467:7316, pp. 687–91, http://www.nature.com/nature/journal/v467/n7316/abs/nature09392.html - supplementary-information. Accessed 26 February 2018.

Morten, T. (2013), *Hyperobjects: Philosophy and Ecology after the End of the World*, Minnesota: University Of Minnesota Press.

Nielsen, K. H. (2008), 'Nanotech, blur and tragedy in recent artworks by Gerhard Richter', *Leonardo,* 41:5, pp. 484–52.

O'Sullivan, S. (2013), 'A diagram of the finite-infinite relation: Towards a Bergsonian production of subjectivity', in J. Ó. Maoilearca and C. d. Mille (eds), *Bergson and the Art of Immanence: Painting, Photography, Film*, Edinburgh: Edinburgh University Press, pp. 165–86.

Panofsky, E. (1944), 'Renaissance and renascences', *The Kenyon Review,* 6:2, pp. 201–36.

——— (1991), *Perspective as Symbolic Form*, New York: Zone Books.

Parkinson, G. (2004), 'Surrealism and quantum mechanics: Dispersal and fragmentation in art, life, and physics', *Science in Context,* 17:04, pp. 557–77.

Paul D. Miller, S. M. (2014), *The Imaginary App*, Cambridge: The MIT Press.

Peirce, C. S. (1892), 'The doctrine of necessity examined', *The Monist,* 2:3, pp. 321–37.

Petrie, B. (1974), 'Boccioni and Bergson', *The Burlington Magazine,* 116:852, pp. 140–47.

Poe, E. A. (2000), *Descent into the Maelstrom*, Charlottesville: University of Virginia Library.

Poincare, H. (1905), *Science and Hypothesis*, London: Walter Scott Publishing.

Pound, E. and Fenollosa, E. (1967), *Instigations of Ezra Pound*, Freeport: Books for Libraries Press.

Redefining the Kilogram: Silicon (10 November 2014), http://www.nist.gov/pml/si-redef/kg_new_silicon.cfm. Accessed 26 February 2018.

Reynolds, N. (2001), 'Turner prize won by man who turns lights off', *The Telegraph*, http://www.telegraph.co.uk/news/uknews/1364860/Turner-Prize-won-by-man-who-turns-lights-off.html. Accessed 26 February 2018.

Russell, F. (2012), 'Graphite: Traces of the past, shades of the future', in R. Collis (ed.), *Graphite*, London: GV Art.

Schrenck-Notzing, A. v. (1923), *Phenomena of Materialisation* (trans. E. E. F. d'Able), London: Kegan Paul, Trench, Trubner and Co. Ltd.

Serres, M. (1982), *Hermes: Literature, Science, Philosophy* (eds J. V. Harari and D. F. Bell), Baltimore: John Hopkins University Press.

——— (2000), *The Birth of Physics* (trans. J. Hawkes), Manchester: Manchester Clinamen Press Limited.

Smith, D. (2013), 'Quantum entanglement and quantum computing | Caltech', http://www.caltech.edu/news/quantum-entanglement-and-quantum-computing-39090. Accessed 26 February 2018.

Susskind, L. (2013), *Why is Time a One-Way Street?* (Video), Santa Fe: Santa Fe Institute.

Sylvester, D. (1987), *The Brutality of Fact: Interviews with Francis Bacon*, London: Thames & Hudson.

——— (2001), *About Modern Art*, New Haven: Yale University Press.

Tegmark, M. (2015), 'Consciousness as a state of matter', *Chaos Solitons Fractals,* 76, pp. 238–70.

Thomas, G. B. (2003), 'Calculus and analytic geometry', in M. D. Weir, R. L. Finney and G. B. Thomas (eds), *Calculus,* alternate ed., Boston: Addison-Wesley.

Thomas, P. (2006), 'Reconfiguring Space', *doctoral thesis,* University of Western Australia, Perth, http://repository.uwa.edu.au/R/-?func=dbin-jump-full&object_id=6912&silo_library=GEN01. No longer available.

——— (2008), 'Nanoessence: God, the first nano assembler', *Technoetic Arts,* 6:3, pp. 217–31.

——— (2011), 'Atomism: Residual images within silver', *Leonardo Electronic Almanac,* 17:1, pp. 152–55.

——— (2013), *Nanoart: The Immateriality of Art*, Bristol: Intellect Books.

Trust, T. V. S. (Producer) (1979), 'Richard Feynman Video – Part 2: Fits of reflection and transmission – Quantum behaviour', *The Douglas Robb Memorial Lectures*, http://www.vega.org.uk/video/programme/46. Accessed 26 February 2018.

Tuma, K. A. (2002), 'Cézanne and Lucretius at the red rock', *Representations,* 78:1, pp. 56–85.

Turok, N. (2012), 'The universe within: From quantum to cosmos', *CBC Massey Lectures.*

Vaughan, M. (2007), 'Introduction: Henri Bergson's "Creative Evolution"', *SubStance,* 36:114, pp. 7–24.

Virilio, P. (2011), *The Original Accident*, Cambridge: Polity.

Wilson, S. (2001), *Information Arts: Intersections of Art, Science, and Technology*, Cambridge: MIT Press.

Wittgenstein, L. (1961), *Tractatus Logico- Philosophicus* (trans. D. F. Pears and B. F. McGuiness), London: Routledge & Kegan Paul.

Youngs, I. (2015), 'Art and science collide to reopen Whitworth gallery', http://www.bbc.com/news/entertainment-arts-31006334. Accessed 26 February 2018.

Żurek, W. H. (2009), 'Quantum Darwinism', *Nature Physics,* 5:3, pp. 181–88.

Author's Biography

Paul Thomas is a Professor of Fine Art at UNSW Art and Design. He is the Director of the Studio for Transdisciplinary Art Research (STAR) and is the instigator of the Transdisciplinary Imaging Conference series 2010-2018. In 2000, he instigated and was the founding Director of the Biennale of Electronic Arts Perth 2002, 2004 and 2007. Thomas is a pioneer of transdisciplinary art practice. His current practice led research takes not only inspiration from nanoscience and quantum theory, but actually operates there. Thomas's current research 'Quantum Consciousness' is based on the experiments being conducted by Professor Andrea Morello, Quantum Nanosystems, UNSW, looking at immersing the viewer in a visualization and sonification of quantum phenomena in the development of quantum computing. He has exhibited nationally and internationally with his previous nano artworks *Mulitverse* based on Richard Feynman's diagrams of photons reflecting from a mirror, *Nanoessence* which explored the space between life and death at a nano level and *Midas* which researched what is transferred when skin touches gold at a nano level. His current publications are, *Nanoart: The Immateriality of Art*, *Reconfiguring Space, Relive Media art Histories*, co-edited with Sean Cubitt, *Interference Strategies* and *Cloud and Molecular Aesthetics* co-edited with Lanfranco Aceti and Edward Colless.

Index